Emerging Pathways of Vaccine Adjuvants

Scrivener Publishing
100 Cummings Center, Suite 541J
Beverly, MA 01915-6106

Publishers at Scrivener
Martin Scrivener (martin@scrivenerpublishing.com)
Phillip Carmical (pcarmical@scrivenerpublishing.com)

Emerging Pathways of Vaccine Adjuvants

A Nonspecific Stimulant of the Immune System

Edited by

Vivek P. Chavda

Dept. of Pharmaceutics and Pharmaceutical Technology, LM College of Pharmacy, Ahmedabad, Gujarat, India

and

Vasso Apostolopoulos

School of Health and Biomedical Sciences, RMIT University, Melbourne VIC, Australia

Scrivener Publishing

WILEY

This edition first published 2025 by John Wiley & Sons, Inc., 111 River Street, Hoboken, NJ 07030, USA and Scrivener Publishing LLC, 100 Cummings Center, Suite 541J, Beverly, MA 01915, USA

For more information about Scrivener publications please visit www.scrivenerpublishing.com.

Wiley Global Headquarters
111 River Street, Hoboken, NJ 07030, USA

For details of our global editorial offices, customer services, and more information about Wiley products visit us at www.wiley.com.

Library of Congress Cataloging-in-Publication Data

ISBN 978-1-394-23761-6

Front cover image courtesy of Adobe Firefly
Cover design by Russell Richardson

Set in size of 11pt and Minion Pro by Manila Typesetting Company, Makati, Philippines

Printed in the USA

10 9 8 7 6 5 4 3 2 1

Contents

Preface

Adjuvants play a crucial role in vaccine formulations by boosting the immunogenicity of antigens, thereby enhancing vaccine efficacy. While antigens can initiate immune responses independently, adjuvants amplify these responses. They do so by stimulating antigen-presenting cells and facilitating the maturation of these cells to effectively present antigenic peptides to both T and B cells. The main objective of this book is to provide readers with an in-depth understanding of the latest advancements in adjuvant technology. Ultimately, the book aims to drive progress in vaccine research, paving the way for the development of more potent and safer vaccines to address global health threats.

This book provides a comprehensive overview of the evolving landscape of vaccine adjuvants, encompassing a wide range of topics critical to their design, development, and application. Chapter 1 is an introductory chapter on adjuvants and Chapter 2 presents the cutting-edge field of *in silico* adjuvant design and validation, shedding light on computational approaches to optimize adjuvant properties. The chapter explores the intricate relationship between adjuvants and immunity (Chapter 3), elucidating how these immunomodulators enhance vaccine responses. Novel formulation strategies for vaccines incorporating adjuvants are discussed (Chapter 4), along with detailed characterization methods to ensure their quality and performance. The book also highlights the role of adjuvants in licensed vaccines (Chapter 5), emphasizing their contribution to vaccine efficacy. Emerging nanomaterial-based adjuvants (Chapter 6) and innovative non-invasive routes of vaccine delivery (Chapter 7) are explored as promising avenues for future vaccine development. Regulatory guidelines governing vaccine adjuvants are outlined to navigate the complex landscape of vaccine approval and licensing (Chapter 8). Importantly, the book addresses vaccine safety concerns associated with adjuvants (Chapter 9), discussing strategies to mitigate risks.

Chapter 10 highlights the limitations of adjuvants and explores future directions to advance the field of vaccinology.

The Editors
Vivek P. Chavda
Vasso Apostolopoulos
November 2024

1

Adjuvants Boosting Vaccine Effectiveness

Vasso Apostolopoulos

School of Health and Biomedical Sciences, RMIT University, Melbourne VIC, Australia

Abstract
Vaccine development has evolved significantly with the identification and isolation of specific antigens, leading to subunit vaccines. Adjuvants, crucial in modern vaccine design, enhance antigen immunogenicity, allowing for more effective vaccines that stimulate both humoral and cell-mediated immunity. Conventional adjuvants, including aluminum salts, SAF-1, QS-21, and squalene-based adjuvants such as MF59 and AS03, play pivotal roles in enhancing vaccine efficacy. Particulate adjuvants, including liposomes, immunostimulatory complexes, and emulsions like MF59 and AS03, offer improved antigen stability and targeted delivery. Additionally, immunostimulatory adjuvants like Toll-like receptor agonists, monophosphoryl lipid A, cytokines, and CpG oligodeoxynucleotides directly activate immune responses. Approved adjuvants, AS01, AS03, AS04, MF59, Matrix-M, and virosomes are key adjuvants in approved human vaccines, enhancing immune responses and vaccine efficacy. Despite advancements, ongoing research is required to optimize adjuvant safety and efficacy in order to develop safer and more effective vaccines against infectious diseases and cancers.

Keywords: Adjuvants, vaccination, AS01, MF59, Matrix-M, virosomes, SAF-1, QS-21

1.1 Vaccines Over the Years

The history of vaccination spans over a millennium, with early attempts to prevent infectious diseases dating back to 1000 A.D. in China, where smallpox vesicles were used for inoculation. Edward Jenner's work in

Email: vasso.apostolopoulos@rmit.edu.au

Vivek P. Chavda and Vasso Apostolopoulos (eds.) Emerging Pathways of Vaccine Adjuvants: A Nonspecific Stimulant of the Immune System, (1–14) © 2025 Scrivener Publishing LLC

the late 1700s marked a significant advancement when he observed that individuals who had contracted cowpox were protected against smallpox. By 1796, Jenner successfully immunized a young boy with cowpox, confirming protection against smallpox. Louis Pasteur furthered the field by demonstrating the use of attenuated pathogens as vaccines in the late 19th century. He attenuated *Pasteurella septica* to develop a vaccine against fowl cholera and later applied a similar approach to *Bacillus* anthrax, achieving remarkable success in protecting farm animals. Additionally, Pasteur's work with rabies marked a significant milestone in the development of live virus vaccines. In the realm of dead organism vaccines, the Salk vaccine against poliomyelitis, developed in 1960, had a profound impact on disease incidence before being succeeded by the Sabin vaccine. Challenges persisted in producing killed vaccines due to potential destruction of important antigenic components.

The identification and isolation of specific antigens responsible for protection paved the way for "subunit" and "extract" vaccines. For instance, diphtheria and tetanus toxoids were purified and inactivated using formalin, retaining their antigenicity but reducing adverse reactions. Despite these advancements, the history of vaccine development is not without setbacks. Disasters such as the Lubeck Disaster in 1932, where infants were mistakenly given *Mycobacterium tuberculosis* instead of BCG vaccine, and the Cutter Disaster in 1955, where a faulty polio vaccine led to cases of poliomyelitis, highlighted the need for stringent quality control and safety measures. As public awareness and standards for vaccine safety have increased, modern vaccinology has embraced advancements in genetics, chemistry, peptide synthesis, protein production methods, DNA, mRNA, x-ray crystal structures, molecular biology, and immunology, allowing for the development of safer and more efficient vaccines [1]. However, there are still many obstacles for their clinical use, and the limited immunogenicity of many of these candidates has hindered their development as potential vaccines. Strategies to enhance the immunogenicity of candidate vaccines are therefore critical. As such, adjuvants have been developed to enhance immunogenicity of vaccines, aiming to overcome their limited efficacy. These advancements are critical for optimizing the clinical potential of novel vaccine candidates.

1.2 Adjuvants in the Modern Era

Adjuvants play a pivotal role in modern vaccine development, enhancing the immune response to antigens and thereby improving vaccine efficacy [2–4]. While antigens alone can stimulate the immune system to some extent, adjuvants amplify this response, making vaccines more effective at inducing both humoral and cell-mediated immunity. This is particularly crucial for subunit vaccines, which consist of purified antigens and often require adjuvants to boost their immunogenicity. Additionally, adjuvants can help reduce the amount of antigen needed per dose, which is beneficial for both vaccine production and delivery. Adjuvants enable the use of novel vaccine technologies, such as synthetic peptides and recombinant proteins, which may otherwise lack sufficient immunogenicity to elicit a protective immune response [5]. Despite their importance, the development and use of adjuvants in human vaccines have been limited by safety concerns, requiring the need for rigorous testing and evaluation. As the area of vaccine development continues to advance, the discovery and optimization of safe and effective adjuvants remain a critical area of research, holding the potential to revolutionize vaccine design and contribute to global health by combating infectious diseases more effectively [6].

1.3 Conventional Adjuvants

Adjuvants play a crucial role in enhancing antigen immunogenicity, amplifying both humoral and cell-mediated immune responses. A widely used adjuvant in experimental animals is complete Freund's adjuvant (CFA), a water-in-oil emulsion containing killed *M. tuberculosis*. Despite its effectiveness and long sustained immune responses, CFA is not suitable for human use due to its propensity to induce granulomas, fever, and inflammation. Incomplete Freund's adjuvant, which lacks the mycobacterial component, has been evaluated, which does not induce granulomas and is safer than CFA, but it is still not approved for human vaccines due to other safety concerns. However, aluminum salts approved for human use in the 1930s, being either as aluminum hydroxide or aluminum phosphate, are the most widely used adjuvants in human vaccines. They enhance the immune response by forming a depot at the injection site, facilitating antigen uptake by antigen-presenting cells and stimulating cytokine secretion [7–9]. Alum primarily stimulates humoral immune responses and is used in vaccines against diphtheria, tetanus, and hepatitis B. However, aluminum-based adjuvants primarily stimulate humoral immune responses and are limited in cell-mediated immune stimulation (Figure 1.1). Emerging conventional adjuvants include the following:

a. **SAF-1:** Comprising squalene oil, threonyl-MDP, and non-ionic block polymers, SAF-1's block polymers act as

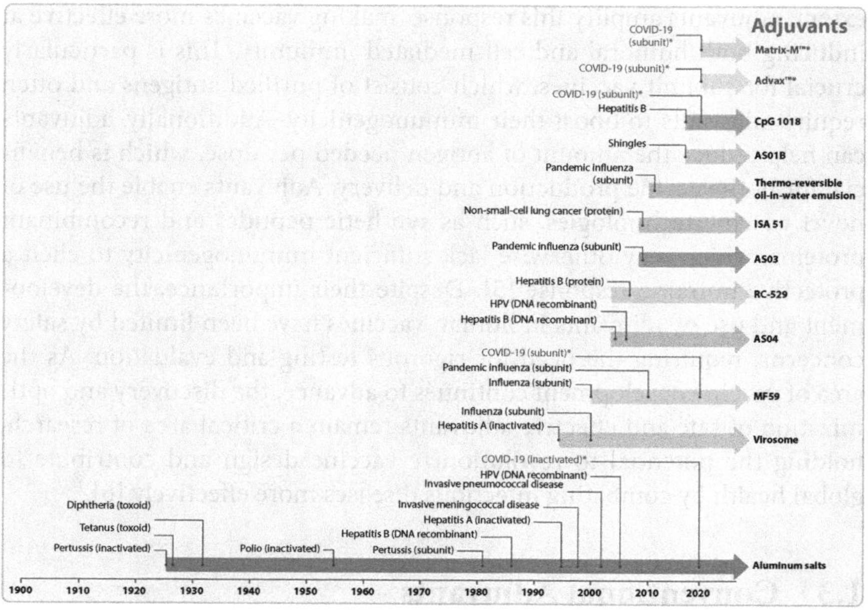

Figure 1.1 A timeline of vaccine adjuvant development over the last 100 years and used in humans. Figure adapted by content experts Iwasaki, A., Lee, J-H. and Omer, S.B at Biorender.com as Figure 1.1 in https://www.cell.com/cell/pdf/S0092-8674(20)31327-X.pdf

adhesive molecules, enhancing antigen presentation and have been used in malaria and influenza vaccine studies [10–12].

b. **QS-21:** QS-21 is a potent vaccine adjuvant sourced by extraction from the Chilean soapbark tree (*Quillaja saponaria*). QS-21 exhibits freeze-thaw stability and has shown promise as an adjuvant for inducing specific CD8+ T-cell responses and exhibits minimal toxicity [13]. Quil A is also derived from *Quillaja saponaria* tree.

c. **Monophosphoryl Lipid A:** A derivative of lipopolysaccharide has been used as an adjuvant in vaccines to enhance antibody and T-cell immune response to antigens. Monophosphoryl lipid A binds to Toll-like receptor 4 (TLR4) on antigen-presenting cells stimulating pro-inflammatory cytokines, maturation, and activation of antigen-presenting cells [14, 15]. AS04 is the best known formulation, which incorporates both monophosphoryl lipid A and aluminum hydroxide.

d. **Ribi Formulation:** Incorporating mycobacterial cell walls and monophosphoryl lipid A, this formulation has demonstrated superior antibody titers and both humoral and cellular immune responses compared to aluminum hydroxide adjuvants [14, 16, 17].

e. **Squalene-Based Adjuvants:** MF59 by Novartis approved in 1997 and AS03 by GlaxoSmithKline approved in 2013 are oil-in-water emulsions containing squalene, a naturally occurring lipid. MF59 and AS03 adjuvants enhance antigen uptake by antigen-presenting cells and stimulate immune cells at the injection site, resulting in activation of both humoral and cell-mediated immune responses. MF59 is used in seasonal influenza vaccines for older adults, whereas AS03 is used in some pre-pandemic (H5N1) and pandemic influenza vaccines [5, 18–20].

f. **Bacterial Toxoids:** Toxoids, such as detoxified forms of tetanus and diphtheria toxins, can serve as adjuvants when co-administered with antigens [21]. They provide T cell help and can enhance the immune response to the co-administered antigen as were shown to be effective in anti-cancer peptide based vaccines [22–27].

g. **Mineral Salts:** Besides aluminum salts, other mineral salts like calcium phosphate and calcium carbonate have been used as adjuvants to stabilize antigens and enhance their immunogenicity. These salts can adsorb antigens and facilitate their uptake by antigen-presenting cells.

Conventional adjuvants have been instrumental in the success of several vaccines by improving their efficacy and durability. However, they often have limitations, such as inducing primarily humoral immune responses or having reactogenicity concerns, which have driven the search for novel adjuvants with improved safety profiles and broader immunostimulatory capabilities.

1.4 Particulate Adjuvants

Particulate adjuvants are a class of adjuvants that consist of particles designed to enhance the immune response to co-administered antigens. These adjuvants are often formulated as nanoparticles, liposomes, or other particulate structures to improve antigen delivery, uptake by antigen-presenting cells and subsequent activation of the immune system. Some notable types of particulate adjuvants include the following:

a. **Liposomes:** Liposomes, phospholipid-based vesicles, have been extensively studied since the 1970s for targeted drug delivery and immunoadjuvant applications [28]. They offer a versatile platform for adjuvant design, with the ability to entrap antigens, cytokines, and other immunomodulators [29]. As such, enhanced immune responses with liposome-entrapped antigens compared to free antigens have been shown, including some of the original studies 30 years ago against influenza virus A/PR/8 envelope proteins [30, 31]. Liposomal-based vaccines hold promise for more effective and tailored approaches in the design of vaccines against infectious diseases, cancer, and other health challenges.
b. **ISCOMs (Immunostimulatory Complexes):** ISCOMs, composed of Quil A adjuvant and peptides, achieve enhanced antigen immunogenicity with reduced adjuvant concentrations. These complexes induce both humoral and cell-mediated immune responses and have demonstrated promise in various animal models, including vaccines against hepatitis B, hepatitis C, influenza virus, malaria, human immunodeficiency, and certain veterinary vaccines [32–37].
c. **Emulsions:** Emulsion-based adjuvants, such as MF59 and AS03, consist of oil-in-water or water-in-oil formulations. They can stabilize antigens, promote their uptake by antigen-presenting cells, and enhance immune responses, particularly in elderly individuals. Indeed, MF59 is used in seasonal influenza vaccines for elderly individuals who have not responded to standard influenza vaccines [19, 38, 39]. In addition, MF59 has been evaluated in vaccines against meningococcus B, SARS-CoV-2, and malaria [40].
d. **Virosomes:** Virosomes are reconstituted viral envelopes devoid of viral genetic material. They can encapsulate antigens and fuse with cell membranes, facilitating antigen delivery and uptake by antigen-presenting cells. Virosomes are used in vaccines like Inflexal V for influenza [41].
e. **Nanoparticles:** Nanoparticles made of biodegradable polymers or inorganic materials can be used to encapsulate antigens and/or adjuvants [42]. These nanoparticles can protect antigens from degradation, target them to specific

cell types, and promote antigen uptake and presentation by antigen-presenting cells [43–45].

f. **Microparticles:** They are larger than nanoparticles but smaller than cells and can be made from various materials, including polymers and proteins. They can be designed to secrete antigens and adjuvants in a controlled manner, enhancing antigen presentation and immune stimulation. Key findings 30 years ago noted that size of the particle was important to stimulate different arms of the immune system [46].
g. **Nanogels:** Nanogels are hydrogel-based nanoparticles, which encapsulate antigens and adjuvants. They provide sustained release of encapsulated components, improve stability, and enhance antigen uptake and presentation by antigen-presenting cells.

Particulate adjuvants offer several advantages, including improved antigen stability, targeted delivery, and enhanced immune stimulation. They are being studied for use in various vaccines against infectious diseases, cancers, and other conditions to improve vaccine efficacy and facilitate the development of novel vaccine formulations.

1.5 Immunostimulatory Adjuvants

Immunostimulatory adjuvants enhance the immune response by directly activating immune cells or signaling pathways. Such examples include the following:

a. **TLR Agonists:** TLR agonists, such as Poly I:C (TLR3) and R848, mimic pathogen-associated molecular patterns to stimulate innate immune responses [47]. In addition, imiquimod, a synthetic imidazoquinolinone compound, binds to TLR7 stimulating pro-inflammatory immune responses. Alum Plus is an improved adjuvant, which combines aluminum salts with TLR agonists.
b. **Monophosphoryl Lipid:** Stimulates immune cells via binding to TLR4.
c. **Cytokines:** Incorporation of cytokines into adjuvants and vaccine formulations has been shown to enhance immune

responses, such as interleukin (IL)-1, IL-2, interferon (IFN)-gamma, and granulocyte-macrophage colony-stimulating factor (GM-CSF).

d. **CpG Oligodeoxynucleotides:** These synthetic DNA sequences stimulate TLR9, promoting a Th1-biased immune response. CpG adjuvants are being investigated for various vaccines, including those against infectious diseases and cancer [48–51].

1.6 Approved Adjuvants for Human Use

Improved adjuvants are essential for enhancing the efficacy of vaccines by boosting the immune response to antigens. While aluminum salts have been key in human vaccines, their limitations, such as variable efficacy and lack of cell-mediated immune stimulation, have directed research into novel adjuvant formulations (Figure 1.1). Several adjuvants have been approved in human vaccines:

a. **AS01:** The adjuvant used in the Shingrix vaccine against shingles is a combination of monophosphoryl lipid A and QS21 inducing robust immune responses even in older adults [52–54].

b. **AS03:** AS03 adjuvant is an oil-in-water emulsion based on squalene, a natural lipid, polysorbate 80, and a-tocopherol, a form of vitamin E. It was approved in 1997 by Novartis to be used in influenza vaccines. AS03 was also present in the Pandemrix influenza vaccine of the H1N1 influenza pandemic 2009-2010. Even though Pandemrix showed strong immune responses and protection against H1N1 infections, there were increased risks of narcolepsy. As such, the use of Pandemrix was discontinued in several countries, and alternative vaccines without the AS03 adjuvant were used for subsequent influenza seasonal and pandemic response vaccines.

c. **AS04:** This adjuvant combines aluminum hydroxide with monophosphoryl lipid A, a detoxified derivative of bacterial lipopolysaccharide, which has been shown to enhance both humoral and cell-mediated immune responses especially in human papilloma virus (HPV) vaccines. Indeed, Cervarix vaccine against HPV types 16 and 18; the Gardasil vaccine against HPV types 6, 11, 16, and 18; and Gardasil

9 against HPV types 11, 16, 18, 31, 33, 45, 52, and 58 comprise the adjuvant AS04 [55–58].

d. **MF59:** This is composed of squalene oil, polysorbate 80, and sorbitan trioleate and has been used in seasonal influenza vaccines. MF59 stimulates immune response, particularly in older individuals who have reduced response to standard influenza vaccines.
e. **Matrix-M:** A saponin-based adjuvant, purified from *Quillaja saponaria* Molina tree, combined with cholesterol and phospholipids to form 40-nm–like nanoparticles. Matrix-M enhances Th1 and cellular immune responses to several antigens and has a favorable safety profile. In fact, the Novavax (NVX-CoV2373) COVID-19 vaccine includes matrix-M adjuvant [59].
f. **Virosomes:** These are reconstituted viral envelopes containing no viral genetic material but capable of fusing with cell membranes and antigen uptake. Virosomal adjuvants are used in the influenza vaccine, Inflexal V, to enhance immune responses [41, 60–63].

1.7 Conclusion

The history of vaccination, spanning over a millennium, has witnessed remarkable advancements, from early inoculations in China to the modern era of sophisticated vaccine technologies. Edward Jenner and Louis Pasteur laid the foundation for vaccine development, whereas the identification of specific antigens enabled the creation of subunit and extract vaccines. Despite these strides, challenges such as safety concerns and limited immunogenicity persist, driving the need for innovative solutions like adjuvants. Adjuvants, crucial in modern vaccine development, enhance immune responses, making vaccines more effective and enabling the use of novel technologies. While conventional adjuvants like aluminum salts have been foundational, their limitations have spurred research into safer and more efficient options. Particulate adjuvants, such as liposomes and ISCOMs, offer improved antigen stability and targeted delivery, whereas immunostimulatory adjuvants like TLR agonists and cytokines directly activate immune cells, enhancing vaccine efficacy. Approved adjuvants, AS01, AS03, AS04, Matrix-M, MF59, and virosomes, have revolutionized vaccine formulations, enhancing immune responses against diseases like shingles, influenza, and HPV. Matrix-M™ in the Novavax COVID-19 vaccine exemplifies the potential of innovative adjuvants in pandemic responses. In conclusion, the ongoing evolution of vaccine technologies and adjuvants holds promise for safer and more effective vaccines. In the face of persistent global health challenges, enhancing vaccine design and delivery is essential to ensure the well-being of people worldwide.

References

1. Chavda, V.P. *et al.*, Protein subunit vaccines: Promising frontiers against COVID-19. *J. Control. Release*, 366, 761–782, 2024.
2. Cui, Y. *et al.*, Vaccine adjuvants: current status, research and development, licensing, and future opportunities. *J. Mater. Chem. B*, 12, 4118–4137, 2024.
3. Lavelle, E.C. and McEntee, C.P., Vaccine adjuvants: Tailoring innate recognition to send the right message. *Immunity*, 57, 4, 772–789, 2024.
4. Lin, Y.J., Zimmermann, J., Schulke, S., Novel adjuvants in allergen-specific immunotherapy: where do we stand? *Front. Immunol.*, 15, 1348305, 2024.
5. Huang, Z. *et al.*, Research progress on emulsion vaccine adjuvants. *Heliyon*, 10, 3, e24662, 2024.
6. Greenfield, E.A., Preparing and Using Adjuvants. *Cold Spring Harb. Protoc.*, 2019, 1, 2019.
7. Bo, C. *et al.*, Physicochemical properties and adsorption state of aluminum adjuvants with different processes in vaccines. *Heliyon*, 9, 8, e18800, 2023.
8. Mbhele, Z. *et al.*, Evaluation of Aluminium Hydroxide Nanoparticles as an Efficient Adjuvant to Potentiate the Immune Response against Clostridium botulinum Serotypes C and D Toxoid Vaccines. *Vaccines (Basel)*, 11, 9, 2023.
9. Zhang, T. *et al.*, Research Progress of Aluminum Phosphate Adjuvants and Their Action Mechanisms. *Pharmaceutics*, 15, 6, 1756, 2023.
10. Byars, N.E. *et al.*, Enhancement of antibody responses to influenza B virus haemagglutinin by use of a new adjuvant formulation. *Vaccine*, 8, 1, 49–56, 1990.
11. Kenney, J.S. *et al.*, Influence of adjuvants on the quantity, affinity, isotype and epitope specificity of murine antibodies. *J. Immunol. Methods*, 121, 2, 157–66, 1989.
12. Pye, D. *et al.*, Selection of an adjuvant for vaccination with the malaria antigen, MSA-2. *Vaccine*, 15, 9, 1017–23, 1997.
13. Martin, L.B.B. *et al.*, Complete biosynthesis of the potent vaccine adjuvant QS-21. *Nat. Chem. Biol.*, 20, 4, 493–502, 2024.
14. Alugupalli, K.R., Monophosphoryl Lipid A-based Adjuvant to Promote the Immunogenicity of Multivalent Meningococcal Polysaccharide Conjugate Vaccines. *Immunohorizons*, 8, 4, 317–325, 2024.
15. Verpalen, E. *et al.*, Structure-Based Design and Synthesis of Lipid A Derivatives to Modulate Cytokine Responses. *Chemistry*, e202400429, 2024.

16. Rohokale, R., Guo, J., Guo, Z., Monophosphoryl Lipid A-Rhamnose Conjugates as a New Class of Vaccine Adjuvants. *J. Med. Chem.*, 67, 9, 7458–7469, 2024.
17. Romerio, A. and Peri, F., Cleaner synthesis of preclinically validated vaccine adjuvants. *Front. Chem.*, 11, 1252996, 2023.
18. Bai, S. *et al.*, Comparison of Immunogenicity of Alum and MF59-Like Adjuvant Inactivated SARS-CoV-2 Vaccines Against SARS-CoV-2 Variants in Elderly Mice. *Viral Immunol.*, 36, 8, 526–533, 2023.
19. Shichinohe, S. and Watanabe, T., Advances in Adjuvanted Influenza Vaccines. *Vaccines (Basel)*, 11, 8, 1391, 2023.
20. Xu, K. *et al.*, The Immunogenicity of CpG, MF59-like, and Alum Adjuvant Delta Strain Inactivated SARS-CoV-2 Vaccines in Mice. *Vaccines (Basel)*, 12, 1, 60, 2024.
21. Prygiel, M. *et al.*, Adjuvant Effect of Whole-Cell Pertussis Component on Tetanus Toxoid Potency in Murine Model. *Biomedicines*, 11, 7, 1795, 2023.
22. Apostolopoulos, V. and McKenzie, I.F., Cellular mucins: targets for immunotherapy. *Crit. Rev. Immunol.*, 14, 3-4, 293–309, 1994.
23. Apostolopoulos, V. and McKenzie, I.F.C., Cellular Mucins: Targets for Immunotherapy. *Crit. Rev. Immunol.*, 37, 2-6, 421–437, 2017.
24. Apostolopoulos, V., Pietersz, G.A., McKenzie, I.F., MUC1 and breast cancer. *Curr. Opin. Mol. Ther.*, 1, 1, 98–103, 1999.
25. Apostolopoulos, V. *et al.*, The immunogenicity of MUC1 peptides and fusion protein. *Cancer Lett.*, 90, 1, 21–6, 1995.
26. Wilkinson, B.L. *et al.*, Synthesis and immunological evaluation of self-assembling and self-adjuvanting tricomponent glycopeptide cancer-vaccine candidates. *Chemistry*, 18, 51, 16540–8, 2012.
27. Xing, P. *et al.*, Phase-I study of synthetic muc1 peptides in breast-cancer. *Int. J. Oncol.*, 6, 6, 1283–9, 1995.
28. Gregoriadis, G., Drug entrapment in liposomes. *FEBS Lett.*, 36, 3, 292–6, 1973.
29. Gregoriadis, G., Davis, D., Davies, A., Liposomes as immunological adjuvants: antigen incorporation studies. *Vaccine*, 5, 2, 145–51, 1987.
30. Gregoriadis, G., Liposomes as immunological adjuvants: approaches to immunopotentiation including ligand-mediated targeting to macrophages. *Res. Immunol.*, 143, 2, 178–85, 1992.
31. Gregoriadis, G. *et al.*, Liposomes enhance the immunogenicity of reconstituted influenza virus A/PR/8 envelopes and the formation of protective antibody by influenza virus A/Sichuan/87 (H3N2) surface antigen. *Vaccine*, 10, 11, 747–53, 1992.
32. Browning, M. *et al.*, Incorporation of soluble antigens into ISCOMs: HIV gp120 ISCOMs induce virus neutralizing antibodies. *Vaccine*, 10, 9, 585–90, 1992.
33. Jakobsen, P.H. *et al.*, Isolation and characterization of a soluble antigen complex of Plasmodium falciparum with pyrogenic properties. *APMIS*, 99, 1, 21–9, 1991.

34. Morein, B., Hu, K.F., Abusugra, I., Current status and potential application of ISCOMs in veterinary medicine. *Adv. Drug Deliv. Rev.*, 56, 10, 1367–82, 2004.
35. Pandey, R.S. and Dixit, V.K., Evaluation of ISCOMs for immunization against hepatitis B. *Curr. Pharm. Biotechnol.*, 10, 7, 709–16, 2009.
36. Polakos, N.K. *et al.*, Characterization of hepatitis C virus core-specific immune responses primed in rhesus macaques by a nonclassical ISCOM vaccine. *J. Immunol.*, 166, 5, 3589–98, 2001.
37. Rivera-Patron, M. *et al.*, ISCOM-like Nanoparticles Formulated with Quillaja brasiliensis Saponins Are Promising Adjuvants for Seasonal Influenza Vaccines. *Vaccines (Basel)*, 9, 11, 1350, 2021.
38. Nguyen, V.H. *et al.*, Cost-effectiveness of introducing an MF59-adjuvanted trivalent influenza vaccine for older adults in Argentina. *Vaccine*, 38, 20, 3682–3689, 2020.
39. Wagner, A. and Weinberger, B., Vaccines to Prevent Infectious Diseases in the Older Population: Immunological Challenges and Future Perspectives. *Front. Immunol.*, 11, 717, 2020.
40. Chavda, V.P. *et al.*, A global picture: therapeutic perspectives for COVID-19. *Immunotherapy*, 14, 5, 351–371, 2022.
41. Levin, Y. *et al.*, A phase 1, open-label, randomized study to compare the immunogenicity and safety of different administration routes and doses of virosomal influenza vaccine in elderly. *Vaccine*, 34, 44, 5262–5272, 2016.
42. Rashidi, N. *et al.*, Nanoparticles in cancer diagnosis and treatment: progress, challenges and opportunities. *J. Drug Deliv. Sci. Technol.*, 95, 105599–105618, 2024.
43. Chavda, V.P. and Apostolopoulos, V., Mosaic receptor binding domain nanoparticles: towards fourth-generation vaccination. *Nanomed. (Lond.)*, 18, 19, 1223–1226, 2023.
44. Minigo, G. *et al.*, Poly-L-lysine-coated nanoparticles: a potent delivery system to enhance DNA vaccine efficacy. *Vaccine*, 25, 7, 1316–27, 2007.
45. Xiang, S.D. *et al.*, Delivery of DNA vaccines: an overview on the use of biodegradable polymeric and magnetic nanoparticles. *Wiley Interdiscip. Rev. Nanomed. Nanobiotechnol.*, 2, 3, 205–18, 2010.
46. Xiang, S.D. *et al.*, Pathogen recognition and development of particulate vaccines: does size matter? *Methods*, 40, 1, 1–9, 2006.
47. Tandel, N. *et al.*, Poly(I:C) and R848 ligands show better adjuvanticity to induce B and T cell responses against the antigen(s). *Heliyon*, 10, 5, e26887, 2024.
48. Chavda, V.P., Pandya, R., Apostolopoulos, V., DNA vaccines for SARS-CoV-2: toward third-generation vaccination era. *Expert Rev. Vaccines*, 20, 12, 1549–1560, 2021.
49. Liang, Z. *et al.*, The orientation of CpG conjugation on aluminum oxyhydroxide nanoparticles determines the immunostimulatory effects of combination adjuvants. *Biomaterials*, 308, 122569, 2024.

50. Pan, H. *et al.*, Orchestrated codelivery of peptide antigen and adjuvant to antigen-presenting cells by using an engineered chimeric peptide enhances antitumor T-cell immunity. *Cancer Immunol. Res.*, 12, 7, 905–920, 2024.
51. Tang, C.K. *et al.*, Molecular basis of improved immunogenicity in DNA vaccination mediated by a mannan based carrier. *Biomaterials*, 30, 7, 1389–400, 2009.
52. Gomez Marco, J.J. *et al.*, [Vaccination against the herpes zoster virus]. *Aten Primaria*, 55, 10, 102710, 2023.
53. Leidner, A.J. *et al.*, Cost-Effectiveness Analysis of Vaccination With Recombinant Zoster Vaccine Among Hematopoietic Cell Transplant Recipients and Persons With Other Immunocompromising Conditions Aged 19 to 49 Years. *Value Health*, 26, 2, 204–215, 2023.
54. Pan, C.X., Lee, M.S., Nambudiri, V.E., Global herpes zoster incidence, burden of disease, and vaccine availability: a narrative review. *Ther. Adv. Vaccines Immunother.*, 10, 25151355221084535, 2022.
55. Einstein, M.H. *et al.*, Comparative humoral and cellular immunogenicity and safety of human papillomavirus (HPV)-16/18 AS04-adjuvanted vaccine and HPV-6/11/16/18 vaccine in healthy women aged 18-45 years: follow-up through Month 48 in a Phase III randomized study. *Hum. Vaccines Immunother.*, 10, 12, 3455–65, 2014.
56. Einstein, M.H. *et al.*, Comparison of long-term immunogenicity and safety of human papillomavirus (HPV)-16/18 AS04-adjuvanted vaccine and HPV-6/11/16/18 vaccine in healthy women aged 18-45 years: end-of-study analysis of a Phase III randomized trial. *Hum. Vaccines Immunother.*, 10, 12, 3435–45, 2014.
57. Germar, M.J. *et al.*, Cost-effectiveness analysis of AS04-adjuvanted human papillomavirus 16/18 vaccine compared with human papillomavirus 6/11/16/18 vaccine in the Philippines, with the new 2-dose schedule. *Hum. Vaccines Immunother.*, 13, 5, 1158–1166, 2017.
58. Schwarz, T.F., Clinical update of the AS04-adjuvanted human papillomavirus-16/18 cervical cancer vaccine. *Cervarix. Adv. Ther.*, 26, 11, 983–98, 2009.
59. Stertman, L. *et al.*, The Matrix-M adjuvant: A critical component of vaccines for the 21(st) century. *Hum. Vaccines Immunother.*, 19, 1, 2189885, 2023.
60. Ansaldi, F. *et al.*, Head-to-head comparison of an intradermal and a virosome influenza vaccine in patients over the age of 60: evaluation of immunogenicity, cross-protection, safety and tolerability. *Hum. Vaccines Immunother.*, 9, 3, 591–8, 2013.
61. Esposito, S. *et al.*, The immunogenicity and safety of a single 0.5mL dose of virosomal subunit influenza vaccine administered to unprimed children aged >/=6 to <36 months: data from a randomized, Phase III study. *Vaccine*, 30, 49, 7005–12, 2012.
62. Herzog, C. *et al.*, Eleven years of Inflexal V-a virosomal adjuvanted influenza vaccine. *Vaccine*, 27, 33, 4381–7, 2009.
63. Mischler, R. and Metcalfe, I.C., Inflexal V a trivalent virosome subunit influenza vaccine: production. *Vaccine*, 20, Suppl 5, B17–23, 2002.

2

In Silico Adjuvant Design and Validation for Vaccines

Vivek P. Chavda[1]*, Anjali P. Bedse[2], Pankti C. Balar[3], Bedanta Bhattacharjee[4], Shilpa S. Raut[2] and Vasso Apostolopoulos[5†]

[1]*Department of Pharmaceutics and Pharmaceutical Technology, LM College of Pharmacy, Ahmedabad, Gujarat, India*
[2]*Department of Pharmaceutics, K K Wagh College of Pharmacy, Nashik, Affiliated to Dr. Babasaheb Ambedkar Technological University, Lonere, Raigad, Maharashtra, India*
[3]*Pharmacy Section, LM College of Pharmacy, Ahmedabad, Gujarat, India*
[4]*Department of Pharmacology and Toxicology, Girijananda Chowdhury Institute of Pharmaceutical Science-Tezpur, Sonitpur, Assam, India*
[5]*School of Health and Biomedical Sciences, RMIT University, Melbourne VIC, Australia*

Abstract
Vaccines are pivotal in reducing disease burden and mortality, particularly in vulnerable populations like children, the immunocompromized, and the elderly. While vaccines provide individual protection, they also establish herd immunity, protecting the broader community. In an attempt to increase immunogenicity of vaccines, adjuvants are designed and incorporated into vaccines. Adjuvants amplify immune responses through various mechanisms, enhancing antigen presentation and immune cell stimulation. Traditional adjuvant discovery is time-consuming, involving target receptor identification, high-throughput screening, and stringent safety evaluations, with predicting human reactogenicity remaining a challenge. In contrast, *in silico* methods, employing computational models, artificial intelligence, and machine learning, expedite adjuvant discovery by predicting efficacy and safety. Immunoinformatics tools, including epitope prediction for B-cell and T-cell recognition and molecular docking simulations, are instrumental in these computational approaches. Next-generation sequencing and high-content

**Corresponding author*: Vivek.chavda@lmcp.ac.in
†*Corresponding author*: vasso.apostolopoulos@rmit.edu.au

Vivek P. Chavda and Vasso Apostolopoulos (eds.) Emerging Pathways of Vaccine Adjuvants: A Nonspecific Stimulant of the Immune System, (15–38) © 2025 Scrivener Publishing LLC

imaging further streamline data analysis, offering insights into adjuvant mechanisms and interactions. A multidisciplinary approach combining computational research, *in vitro* characterization, animal model evaluations, and toxicology studies is crucial. These partnerships validate, refine, and optimize *in silico* predictions, enhancing the predictive accuracy and applicability of computational methods in adjuvant discovery and design.

Keywords: Adjuvant selection, *in-silico*, vaccination, epitope, immunology

2.1 Introduction

2.1.1 Importance of Vaccines and Adjuvants in Immunology

Vaccines are useful to reduce the morbidity and mortality from serious illnesses, particularly those that influence children, immune compromised, and elderly. The majority of individuals who get vaccinated are primarily focused on their immediate protection. From a public health perspective, the benefit of vaccination is the establishment of herd immunity, which helps protect the community at large [1, 2]. Adjuvants boost the immune response to antigens through mechanisms like targeting specific cells, immunomodulation, stimulation of T cells, and depot formation. They can act through multiple mechanisms to enhance immunity against antigens. Adjuvants stimulate receptors such as Toll-like receptors (TLRs), nucleotide binding oligomerization (NOD)-like receptors, and retinoic acid-inducible gene (RIG)-I–like receptors and can stimulate cytokines and chemokines to skew immune responses toward specific pathways. For example, Interleukein (IL)-12 promotes Th1 responses, whereas IL-4 promotes Th2 responses. Adjuvants can also be designed to enhance antigen uptake and presentation by antigen-presenting cells (APCs), such as dendritic cells. Adjuvants, together with small molecules or biologics, directly influence immune cell activity, including immune checkpoint inhibitors or regulatory T-cell agonists [3, 4].

2.1.2 Limitations of Traditional Adjuvant Discovery Methods

The process of adjuvant discovery is labor-intensive. First, a target pattern recognition receptor, like TLR7/8, is identified in the standard adjuvant discovery. Millions of molecules are screened for their ability to agonize the target receptor using a process known as high-throughput screening. The top "hits" of activity are determined to be lead compounds after numerous iterative screening rounds. To optimize these lead compounds, modifications can be made based on structure–activity relationships.

Before advancing to thorough testing, potential lead compounds are combined with antigens in vaccines. Both adjuvants and vaccines are subject to rigorous safety standards. As such, before human clinical trials can begin, extensive *in vivo* studies using mice and non-human primates are essential. Despite advancements in adjuvant development, accurately predicting human reactogenicity and tolerability while achieving the desired immune response remains challenging [5].

2.1.3 Introduction to *In Silico* Approaches for Adjuvant Design

In silico experiments, using computational models, can accelerate adjuvant discovery, often enhanced by machine learning (ML) for predictive algorithms. *In silico* protein modeling has identified adjuvants and used in several vaccines including those against *Plasmodium yoelii* and *Mycobacterium tuberculosis*. Machine learning techniques like black-box optimization, which included active learning, reinforcement learning, and Bayesian optimization, are used in high-throughput adjuvant screening to predict specific characteristics of adjuvants. These properties could include efficacy, safety, and interactions with immune cells. Technological advances, including next-generation sequencing, omics techniques, and ML, enable comprehensive data collection for understanding and evaluating adjuvant efficacy. High-content imaging uncovers hidden patterns, facilitating *in silico* binding simulations and streamlining adjuvant development at all design stages [5].

Databases and *in silico* tools work together in vaccine design. Databases store verified vaccines and components, setting standards for prediction programs, whereas *in silico* tools provide computational methods for designing new vaccines. VIOLIN, a comprehensive web-based vaccine database, includes Vaxign, a genome-based design program [6, 7]. A web-based database, Växjö, collects and analyzes information on vaccine adjuvants, including names, components, structure, and usage details. It connects these with 380 vaccines targeting 81 conditions and over 100 selected adjuvants [8].

2.2 *In Silico* Techniques for Adjuvant Discovery

In silico techniques for adjuvant discovery utilize computational methods to accelerate the identification and design of potential adjuvants, enhancing the efficiency and precision of vaccine development. Main features include the following:

a. Structure retrieval and preparation: All bound ligands are removed from the target protein during pre-processing by using the Protein Data Bank (PDB) database. Eliminating them leads to a simpler computing process and an optimal posture search [9].
b. Preparation of monoclonal antibody: The PDB database is used to obtain antibodies. In order to identify compounds with comparable structural orientation and possible actions, it is used as a ligand in the virtual screening of natural chemicals database [9].
c. Pharmacophore-based virtual screening: A limited number of hits are identified through virtual screening when 3D structures are uploaded to the PDB due to certain stereochemical features.
d. Druglikeness test: This is used to assess a compound's suitability for use as a medication, helping identify compounds with favorable therapeutic properties [9].
e. Molecular dynamics simulation (MDS): Using MDS, docking results are verified, and the behavior of final candidate adjuvants within the protein binding pocket are investigated.
f. *In vitro* validation of candidates: *In vitro* validation has become a standard approach in design and discovery processes following MDS [9].

2.2.1 Immunoinformatics Tools for Epitope Prediction

Immunoinformatics combines immunology with computational biology to analyze and predict immune system interactions, aiding in vaccine design, immunotherapy development, and understanding immune responses at a molecular level. High-throughput methods can provide crucial insights into immune cell-peptide interactions [10]. Immunoinformatics use bioinformatics to analyze immune system characteristics and predict B- and T-cell epitopes [11, 12]. This approach identifies antigenic regions of proteins [13] and offers a cost-effective and time-efficient alternative to laboratory based methods [14, 15].

2.2.1.1 Identification of B-Cell and T-Cell Epitopes

Antigenic peptides are presented by major histocompatibility complex [MHC class I (MHC-I) and MHC class II (MHC-II)] molecules, commonly referred to as human leukocyte antigen (HLA) in humans, for T-cell stimulation. CD8+ T cells recognize peptide–MHC-I complex and CD4+ T cells recognize peptide–MHC-II complex [16]. Using computational tools, immunoinformatics can rapidly predict peptide epitopes presented on MHC molecules, aiding in vaccine design. In contrast, B cells recognize peptides via their cell surface receptor (B-cell receptor), a membrane-bound immunoglobulin. B-cell epitopes can either be continuous (linear) or discontinuous (conformational) types. Computational tools and algorithms are important for accurate epitope prediction [17]. Linear epitopes are consecutive amino acids and do not require the protein to be folded in a 3D conformation, whereas conformational epitopes are amino acids, which are brought together in a 3D conformation in a protein, making their prediction challenging, and, as such, nuclear magnetic resonance (NMR) spectroscopy and/or crystal structures are required. The most antigenic epitopes, however, are conformational epitopes as they mimic the native structure more closely and bind to B-cell receptors with a higher affinity [18].

2.2.1.2 Tools Used for MHC and B-Cell Receptor Binding Prediction Tools

Bioinformatics tools play a crucial role in predicting MHC-I and MHC-II binding epitopes, which are important for understanding CD8+ T-cell immune responses. These tools, including IEDB, NetCTL, MHCPred, NetMHC, nHLAPred, CTLPred, SVMHC, RANKPEP, BIMAS, MAPPP, ProPred, SYFPEITHI, PREDEP, and MHCPEP (for MHC-I), as well as IEDB, NetMHC-II, MHCpred-II, MetaMHC-II, MetaSVMP, Propred-II, RANKPEP, PREDIVAC, SYFPEITH, BIMAS, CTL-Pred, EpiTOP, MHCPEP, EpiVax, PREDEPP, TEPITOPE, EPIPREDICT, EpiDOCK, Concensus, and EpiMatrix (for MHC-II), are instrumental in analyzing peptides' binding potential to MHC molecules. These tools aid in predicting epitopes from various types of antigens, including those from bacteria, viruses, and tumors. Depending on the antigen type, the performance of each program in predicting MHC-II–bound epitopes can vary, highlighting the importance of choosing the appropriate tool for specific antigen analyses [19, 20].

Bepipred, BCpred, ABCpred, Pcipep, etc., are the tools available for predicting B-cell epitopes. The software Bepipred is commonly used to predict B-cell epitopes of viruses, and BCpred is commonly used for bacteria and ABCpred for tumor epitopes [21]. Discotope, Ellipro, CBTope, Epitope, etc., are tools used to predict conformational B-cell epitopes. Each software tool used to predict conformational epitopes of B cells is based on antigen types evaluated across three categories: viral, bacterial, and tumor-specific antigens [22]. All the above tools for both T- and B-cell epitope prediction assists in understanding immune responses mediated by T and B cells and facilitating the design of peptide-based vaccines.

2.2.2 Molecular Docking Simulation

The technique known as "molecular docking" evaluates how molecules align and conform into the binding site of a macromolecular target. Possibilities are generated by searching algorithms and ranked using scoring formulas. Over the past few decades, several softwares have been developed; some of the more well-known ones include AutoDock, AutoDock Vina, DockThor, GOLD, FlexX, and Molegro Virtual Docker. Obtaining the target structure, typically a large biological molecule (protein, DNA, or RNA) is the first step in conducting a docking calculation [23]. These macromolecular structures are readily accessed from the PDB, which offers access to three-dimensional atomic coordinates derived by experimental

Table 2.1 A list of the docking software and scoring function.

Software	Scoring	Posing	Reference
Molegro/ MolDock	Semi-empirical	Alternatively Simplex Evolution and Iterated Simplex	[25]
AutoDock4	Semi-empirical	Algorithm, Genetic Algorithm, or Simulated Annealing	[26]
AutoDock Vina	Empirical/ knowledge-based	Iterated Local Search + BFGS Local Optimizer	[26, 27]
SMINA	Empirical	Monte Carlo stochastic sampling + local optimization	[27]
ICM	Physics-based	Biased Probability Monte Carlo + Local Optimization	[26]
FRED	Empirical (defaults to Chemgauss3)	Conformer generation + Systematic rigid-body search	[28]

techniques, such as x-ray crystal structures [24]. Table 2.1 highlights the docking software and its scoring function.

2.2.2.1 Application of Molecular Docking in Adjuvant Design

Molecular docking plays an important role in adjuvant design by determining the interactions between potential adjuvant molecules and target receptors. This computational approach aids in identifying and optimizing adjuvants that can enhance immune responses, thereby improving vaccine efficacy and design. Some key features include the following:

a. Hit Identification/Virtual Screening: By forecasting the binding affinity of small compounds to a protein or receptor of interest, it aids in the identification of possible therapeutic candidates [27].
b. Lead Optimization: Molecular docking can be utilized to refine the structure of the lead molecule to increase its binding affinity and selectivity after a hit compound has been identified [29].
c. Bioremediation: In bioremediation, molecular docking is used to determine the molecule's binding affinity to enzymes that break down environmental contaminants [30].
d. ADMET Prediction: The absorption, distribution, metabolism, excretion, and toxicity (ADMET) characteristics of small compounds can also be predicted *via* docking [31].
e. Molecular Dynamics Simulation: Molecular docking and molecular dynamic simulations can be used together to determine the dynamic properties of protein–ligand complexes [32].
f. Structure Elucidation: Proteins with unclear structures can also have their structures clarified through the use of molecular docking [32].

2.2.3 Artificial Intelligence and Machine Learning for Adjuvant Design

Several fields of science, including medicine, are fast changing as a result of artificial intelligence (AI) and machine learning (ML). Whereas ML is a subset of AI, ML is the process of using data to create predictions or classifications, either with or without human supervision. AI is the construction of robots or systems that can replicate human thought and conduct. The advent of high performance computers in recent years has expedited the growth in these sectors. Digitalized medical fields like imaging are well-suited to be early users of AI and ML. Digital space facilitates the operation of the imaging pipeline, which includes picture acquisition, reconstruction, interpretation, reporting, and result sharing. This enables the efficient capture of data for AI and ML [33].

2.2.3.1 *Leveraging Large Datasets for Adjuvant Discovery*

A volumetric chest CT scan containing a tumor, like a lung nodule, serves as a valuable dataset for radiomics analysis. The process typically involves several key steps:

a. Identifying the tumor within the scan.
b. Annotating the tumor with semantic attributes often done by expert radiologists [34].
c. Drawing a contour or segmenting the tumor to define its boundaries [35].
d. Calculating various predefined tumor characteristics such as size, mean intensity, image texture, shape, and margin sharpness [36, 37].
e. Utilizing automated learning algorithms to extract relevant features for the specific analysis.
f. Developing a classifier that uses the extracted features to predict clinical outcomes, such as the likelihood of a specific gene mutation, response to treatment, or overall survival [38].

2.2.3.2 *Types of Machine Learning Algorithms Used*

Any ML algorithm's likelihood of success depends on the availability of data, the processing capacity of the machine, and further algorithm improvements. With smaller datasets, traditional ML methods like Naive Bayes, logistic regression, decision trees, and support vector machines are frequently used. The choice of ML technique may also rely on the size of the data. More sophisticated ML models, like convolutional neural networks (CNNs), which are excellent at learning directly from images, might be better for larger datasets, despite the fact that they require more processing power. A subset of ML techniques based on artificial neural networks,

CNNs are an example of deep learning. Artificial neural networks mimic the connectivity of neurons to solve issues by modeling how neurons are arranged in the brain. On either end of the spectrum, ML algorithms can be either supervised or unsupervised.

2.2.3.3 Case Study

In cancer imaging research, a crucial process involves obtaining patient images and preparing them for analysis using ML algorithms. This preparation step ensures that the data conforms to specific standards, including pixel size and image section thickness, which is essential for accurate analysis. The data may include features identified by radiologists or statistically generated features from radiomics. ML algorithms are then applied to these pre-processed images to learn and map the relationships between the imaging data and various clinical or scientific observations. This learning process involves developing mathematical functions, ranging from simple to complex, that are associated with the desired target or output. To develop ML algorithms, researchers often use training datasets, refine them with validation datasets, and finally assess their performance using independent test datasets, preferably from different institutions. These datasets help in training the algorithms, optimizing their performance, and evaluating their accuracy and reliability. Regression models like Lasso, Ridge, ElasticNet, Linear, Cox (Proportional Hazards), and Regression Trees are commonly utilized in analyzing continuous variables within this context. These models aid in understanding and predicting outcomes based on the imaging data, contributing significantly to advancements in cancer imaging and diagnosis [39]. For discrete variables, one can use a variety of classification models, including Bagging, Naive Bays, Support Vector Machines, Decision Trees, Random Forests, KNN (k-nearest neighbors), and Generalized Linear Models [40]. These models can help with therapy response, disease outcomes, characterization and stratification of diseases, and cancer diagnosis [38].

2.2.4 *In Silico* Toxicology Prediction

Predicting chemical toxicity using quantitative structure–activity relationships (QSARs), computational modeling, and algorithm prediction using toxicity data is the aim of *in silico* toxicity. It is regarded as one of the alternatives to animal testing, which generally makes use of a range of computational methods that establish a relationship between a chemical's structure and efficacy or toxicity. The toxicity of various substances, including mixtures, nanomaterials, and chemicals, may be predicted both quantitatively and qualitatively thanks to computational approaches. Each problem has a different set of strengths and limitations related to scope and interpretation; therefore it is critical to choose the most effective approach for the given situation. When chemicals combine with other substances, they can become hazardous, and risk assessments may not account for these side effects. For instance, while combining different metals raises the toxicity of lead and provides an accumulative risk assessment, testing every possible combination of these elements is impractical and results in a lack of raw data regarding the mixture's toxicity [41].

2.2.4.1 *Minimizing Safety Concerns During Adjuvant Design*

Many programs and online platforms can predict chemical toxicity prior to synthesis. Drug design software like Discovery Studio and Pipeline Pilot incorporate toxicology prediction algorithms to identify potentially toxic compounds. While these integrated tools offer valuable insights, several commercial programs, such as Lhasa Derek, Leadscope, and ADMET Predictor, provide more diverse and focused toxicity assessments. Academics use free software or web servers due to their flexibility in developing advanced models across diverse applications, including drug discovery. OCED Toolbox is a recognized toolkit for QSTR-based toxicity modeling. Web servers are user friendly to operate, which is why medicinal chemists and other non-experts in computational toxicology use them. Lazar is one such program that uses a chemical structure sketching user interface to predict toxicity outcomes [42]. ToxTree is an open-source tool that uses a decision tree method to assess toxic hazards [43].

2.2.4.2 *Software Tools for Virtual Toxicity Assessment*

Virtual toxicity assessment software tools use predictive algorithms and advanced modeling to anticipate potential risks of chemical compounds. These tools aid in rapidly identifying safe and effective compounds, accelerating drug development and ensuring consumer safety. Some software tools include:

a. Deductive estimate of risk from existing knowledge (Derek): Derek is a specialized QSAR model developed by Lhasa Ltd., designed to predict toxicity based on established professional guidelines. It utilizes a knowledge-based approach to forecast various toxicity indicators, including genotoxicity, by analyzing chemical structures. Derek categorizes compounds based on their structural features and can also predict the structure–activity relationship in biological systems [44].

b. EPA toxicity estimation software tool (TEST): The Chemistry Development Kit (CDK) Java open-source platform employs eight distinct QSAR methodologies to validate research findings and facilitate the integration of chemical data [45].
c. Danish QSAR: This model, which is repository-based and compatible with 200 QSAR models, contains data on over 600,000 compounds. Physical and chemical properties, acute toxicity, skin corrosion/irritation, and environmental toxicity are the range of predictions [46].
d. VegaHub QSAR: This supports the read-across approach and allows for the simultaneous batch prediction of many compounds by holding 40,000 different types of chemical data and providing prediction results optimized for REACH criteria [47].
e. Toxtree: The program that puts Cramer's decision tree (DT) theory into practice is called Toxtree. Cramer DT is categorized into three groups based on the compound's metabolism, toxicity information, and whether or not it is included in traditional foods [47].
f. PreADMET: Both genotoxicity and carcinogenicity prediction models are available in the PreADMET package [47].

2.3 Case Studies: Successful Applications of *In Silico* Adjuvant Design

2.3.1 Designing Adjuvants Targeting Specific Immune Pathways (e.g., TLR Agonists)

Designing adjuvants targeting specific immune pathways requires a deep understanding of the immune system and the mechanisms of action of adjuvant molecules. The process involves identifying the target immune pathway, understanding the immune pathway's molecular and cellular mechanisms, selecting adjuvant molecules, formulating delivery systems, evaluating formulations in preclinical testing, advancing adjuvant formulations to clinical trials, and obtaining regulatory approval for commercial use.

Collaboration between immunologists, biochemists, pharmacologists, and clinicians is crucial for successful development of adjuvants targeting specific immune pathways. The process includes evaluating the adjuvant formulations in preclinical models, advancing them to clinical trials, and obtaining regulatory approval for commercial use.

One class of pattern recognition receptors known as TLRs is responsible for initiating the host's response to infection by identifying microbial invaders. Strong and rapid innate immune responses are elicited by TLR activation, and these responses create the way for later adaptive immunological responses. Consequently, these TLRs are excellent targets for the creation of useful adjuvants. TLR agonists are being developed for this purpose. Currently, approved vaccines contain TLR agonists like CpG-1018 and monophosphoryl lipid A for their adjuvant function [48]. Immunostimulants are substances that interact with TLRs found on APCs. This interaction results in several effects that boost the adaptive immune response. By engaging with different TLRs, immunostimulants can trigger various pathways, leading to enhanced antigen presentation, increased expression of co-stimulatory signals, and the production of cytokines. TLRs located on the cell surface primarily detect components of microbial membranes. When activated, they induce the secretion of pro-inflammatory cytokines like IL-1β, tumor necrosis factor (TNF)-α, and IL-6. On the other hand, intracellular TLRs, located within vesicles, stimulate the production of type I interferons. These interferons play a crucial role in promoting the differentiation and function of Th1 cells, which are essential for effective immune responses. Specifically, immunostimulants that bind to TLR2 heterodimers initiate signaling via the MyD88 pathway. This activation leads to the activation of nuclear factor-kappa beta (NF-κB), a transcription factor that triggers the production of pro-inflammatory cytokines. Overall, immunostimulants play a vital role in modulating immune responses by engaging with TLRs and influencing downstream signaling pathways [49, 50]. TLR4 and TLR5 initiate signaling via the MyD88 pathway, activating NF-κB and driving naive T cells to Th1-type cell polarization, as shown with mannose conjugated vaccines, which bind to TLR4 and active strong cellular and Th1 responses [51–53]. TLR7/8/9 targets activate NF-κB and interferon regulatory factor 7 (IRF7) via the MyD88 pathway, triggering the production of pro-inflammatory cytokines and type I interferons by APCs [3].

The rapid progression of COVID-19 clinical trials has highlighted the importance of adjuvants based on TLR agonists. These adjuvants have shown significant efficacy, leading to the approval and emergency use authorization of TLR agonist-adjuvanted vaccines against COVID-19.

This success underscores the potential of TLR agonists in enhancing vaccine effectiveness against infectious diseases. Future developments in TLR agonist adjuvants could involve structural modifications, combination therapies, conjugation to antigens, and incorporation into nanoparticle formulations, aiming to create next-generation vaccines. Additionally, there is potential for TLR agonists to be utilized as nasal adjuvants, thereby influencing the immunogenicity of mRNA vaccines and expanding their utility in combating infectious diseases [54].

2.3.2 Development of Multi-Epitope Vaccines with *In Silico* Adjuvant Selection

The selection of immunogenic epitopes is based on their immunogenicity, antigenicity, conservation, and relevance to the disease. Multi-epitope constructs are designed, incorporating epitopes from different antigens or proteins to target various aspects of the pathogen's biology or immune evasion mechanisms. *In silico* adjuvant selection is done using computational methods to analyze the physicochemical properties of potential adjuvants, their ability to enhance immune responses, and their safety profiles. Several algorithms and databases are available for epitope prediction, including tools like NetCTL, IEDB, and Immune Epitope Database. The vaccine formulation is then evaluated in preclinical models, followed by clinical trials and regulatory approval.

Using immunoinformatics and immune simulation analyses, a new multi-epitope vaccine (MEV) against chikungunya virus (CHIKV) structural proteins was developed. This construct proved non-allergenic and immunogenic, effectively triggering immune responses with a suitable synthetic adjuvant. It displayed satisfactory physicochemical properties and accurately predicted B-cell, CD4+, and CD8+ T-cell epitopes. Docking and MD simulations verified the stability of the TLR4-MEV complex [55]. Debarghya Mitra *et al.* developed a MEV using epitope prediction servers and molecular dynamic simulations. The vaccine, involving S1 and S2 spike glycoprotein subunits of SARS-CoV-2 and a chimeric adjuvant, was evaluated for docking with TLR. The study suggested the potential for a vaccine candidate against SARS-CoV-2 [56]. In addition, a filarial MEV using immunoinformatics consisting of linear B-cell and T-cell epitopes was found to be superior in antigenicity compared to current vaccine candidates, suggesting potential cross-protection benefits [57]. Beatriz Meza *et al.* developed a CTB-multiHp vaccine with multiple antigens to enhance *H. pylori* infection protection. The vaccine involved various stages of pathogenesis and induced a specific immune response by B- and T-cell multi-epitopes. The chimeric construct was analyzed and shown to exhibit immune stimulation in *Escherichia coli* BL21 [58]. A new vaccine against *H. pylori* was developed using genomics, proteomics, bioinformatics, and reverse vaccinology, screening PLA1 and Omp6 antigens and enhancing immunity with a new construction strategy. The study proposes a novel strategy for designing MEVs against *H. pylori* using reverse vaccinology and structural vaccinology. Candidate antigens PLA1 and Omp6 were screened, and MEVs were constructed using predicted epitopes. Six constructs were evaluated for stability and B-cell epitope exposure. Construct S1 was selected based on its ability to bind TLRs and its expression level in *E. coli* K12 [59]. Researchers have been exploring ways to address Brucellosis, a zoonotic disease affecting both animals and humans. Unlike some other diseases, there are currently no vaccines available for either animal or human Brucellosis. However, a recent study took an innovative approach by using immunoinformatics to identify antigenic epitopes from Brucella's type IV secretion system proteins, VirB8 and VirB10. This study yielded promising results, identifying several epitopes that could be targeted for vaccine development. Specifically, they found two CD8+ T-cell epitopes, nine helper T-cell epitopes, six linear B-cell epitopes, and six conformational B-cell epitopes. These epitopes were then combined to create a MEV. *In silico* testing, which involves computer simulations rather than real-world experiments, indicated that this vaccine formulation was non-allergenic, safe, and capable of stimulating both cellular and humoral immune responses in individuals with Brucellosis infection. This approach represents a step forward in the quest for effective vaccines against Brucellosis for both animals and humans [60].

2.3.3 Repurposing Existing Drugs as Vaccine Adjuvants Through *In Silico* Analysis

Repurposing existing drugs as vaccine adjuvants through *in silico* analysis is a cost-effective and time-efficient method for identifying novel adjuvant candidates. This process involves selecting a database of known pharmacological properties, screening the drug database using computational methods like molecular docking, MDSs, and QSAR modeling, and identifying potential adjuvants based on their binding affinity and interactions with immune receptors or pathways. Validation of these drugs is achieved through *in vitro* and *in vivo* studies, focusing on immune cell activation assays, cytokine profiling, and antigen presentation. Mechanisms of action are investigated, and the optimal formulation and dosing regimen are determined. Preclinical evaluation is conducted to assess the safety, immunogenicity, and efficacy of the adjuvant. Clinical translation is initiated, and regulatory approval is obtained for commercialization. This approach offers a streamlined approach to identify novel immunomodulatory agents that can enhance vaccine efficacy and accelerate the development of new vaccines [3].

Glucocorticoids, like cortisone, interact with glucocorticoid receptors (GRs) to bind to target genes. GR-induced genes, such as glucocorticoid-induced leucine zipper, have anti-inflammatory effects, and it affects cancer cell biology, tumor microenvironment, and immune system. Dexamethasone treatment increased miR-708 expression in breast cancer cell lines, suppressing NF-κB signaling. The response score of Dexamethasone is positively correlated with innate immune cells in the cancer microenvironment, hence possible correlation with anticancer immunity [61]. Immunomodulation of tumor microenvironment is crucial for therapy, with CD4+ FoxP3+ T regulatory cells (Tregs), suppressing effector T cells in melanoma and breast cancer immune checkpoint inhibitor non-responders. Cyclophosphamide, a well-known chemotherapeutic, has been found to effectively inhibit Tregs at a low dose, offering a promising avenue for opportunity of drug repurposing for immunotherapy [62]. Wang *et al.* identified Treg function in tumor infiltrates based on EZH2 protein enhancer. Inhibiting EZH2 enhances anti-cancer immunity, suggesting machine and deep learning methods to identify novel inhibitors of EZH2 for drug repurposing for Treg-driven immunomodulation [63]. Indoleamine 2,3-dioxygenase is a target of interest for immunomodulation, as it catalyzes the degradation of tryptophan in the tumor microenvironment, activating the GCN2 kinase pathway and influencing Treg cell differentiation. It also binds to the aryl hydrocarbon receptor, causing Treg cells to develop a tolerogenic phenotype. The adenosine A2A receptor is another target of interest for immunomodulation [64].

2.4 Challenges and Future Directions of *In Silico* Adjuvant Design

In silico methods have significantly impacted adjuvant design, offering tools to expedite the discovery and optimization of novel adjuvants. However, these computational approaches face challenges that require attention and refinement to enhance their utility in vaccine development.

One notable limitation of current *in silico* methods is their predictive accuracy. While computational models and ML algorithms can predict the efficacy and safety profiles of potential adjuvants, these predictions often require validation through rigorous laboratory and animal experiments. The complex nature of the immune system poses challenges in accurately simulating and predicting interactions between adjuvants and the immune system based solely on computational models. Another challenge is the integration of *in silico* findings with experimental designs and clinical applications. Bridging the gap between computational predictions and biological *in vitro* and *in vivo* validations are essential and requires the development of robust methodologies for linking *in silico* approaches with experimental data. Further, adjuvant prediction through computational methods poses safety issues and, as such, requires thorough testing and validation in animal models prior to regulatory approval for testing in human clinical trials.

In recent years, the use of peptide-based adjuvants and the integration of peptides into vaccines have gained interest, driven by advancements in peptide synthesis, structural biology, and immunology. As research in peptide-based adjuvants continues to evolve, these innovative approaches hold promise for improving vaccine development. Additionally, the identification of non-canonical binding peptides in *in silico* adjuvant design offers a novel approach to diversify potential adjuvant candidates; such examples have been noted in cancer vaccine design [65–68]. Traditional adjuvants primarily target well-characterized immune receptors and pathways, limiting the diversity and scope of immune modulation. Likewise, peptide mimics, a peptide with a completely different sequence mimicking another sequence are important for enhanced responses and algorithm tools currently cannot identify such peptides; in cancer and in multiple sclerosis, peptide mimics have been shown to modulate immune responses [69–72]. By determining peptide mimics, non-canonical binding peptides, and unconventional immune targets, new avenues for adjuvant discovery and design can be developed, potentially leading to more potent and versatile adjuvants.

To address the current challenges and advance the field of *in silico* adjuvant design, there is a need for interdisciplinary and multidisciplinary collaborations between computational research, *in vitro* experimental research, animal models and clinicians. As such, computational predictions are validated, refined, and optimized, improving the predictive accuracy of *in silico* methods in adjuvant discovery and design.

2.5 Conclusion

In conclusion, vaccines and adjuvants are important in immunology, significantly reducing disease-related morbidity and mortality, especially among vulnerable populations like children, the immunocompromized, and the elderly. While vaccines confer individual protection, they also play a pivotal role in establishing herd immunity, protecting the broader community against infectious diseases. Traditional methods of adjuvant discovery have been compromised by challenges, including time-consuming processes such as target receptor identification, high-throughput screening, and rigorous safety evaluations. Despite these efforts, accurately predicting human reactogenicity remains a difficult task, necessitating the need for more efficient and predictive approaches.

In recent years, *in silico* techniques have emerged as powerful tools in adjuvant design and discovery. Leveraging computational models, ML algorithms, and MDSs, these methods expedite the identification and optimization of novel adjuvants by forecasting their efficacy and safety profiles. Molecular docking simulations and AI-driven analyses further enhance our understanding of the interactions between adjuvants and immune receptors, paving the way for the development of more targeted and potent adjuvant formulations. The advent of next-generation sequencing and high-content imaging technologies has revolutionized data acquisition and analysis, allowing for the comprehensive characterization of immune responses and the identification of potential adjuvant candidates. In addition, the availability of large datasets, curated in databases like VIOLIN and Växjö, has proven invaluable in supporting vaccine design by collating verified vaccines and adjuvants, establishing standards, and facilitating computational design strategies. Furthermore, *in silico* toxicology prediction and virtual toxicity assessment software tools have been instrumental in evaluating the safety profiles of adjuvants, ensuring that only the most promising candidates progress to further preclinical and clinical evaluations. Targeting TLRs and repurposing existing drugs as vaccine adjuvants through *in silico* analysis offer innovative strategies to enhance immune responses while minimizing potential side effects.

As such, the integration of *in silico* approaches, advanced computational techniques, and large-scale data analytics has significantly accelerated the pace of adjuvant discovery and design. These innovative methodologies not only streamline the identification of novel adjuvants but also enhance our understanding of immune responses, paving the way for the development of safer, more effective, and targeted vaccine candidates.

References

1. Verma, S.K., Mahajan, P., Singh, N.K., Gupta, A., Aggarwal, R., Rappuoli, R., Johri, A.K., New-Age Vaccine Adjuvants, Their Development, and Future Perspective. *Front. Immunol.*, *14*, 1043109, 2023, doi: 10.3389/fimmu.2023.1043109.
2. Rodrigues, C.M.C. and Plotkin, S.A., Impact of Vaccines; Health, Economic and Social Perspectives. *Front. Microbiol.*, *11*, 1526, 2020, doi: 10.3389/fmicb.2020.01526.
3. Zhao, T., Cai, Y., Jiang, Y., He, X., Wei, Y., Yu, Y., Tian, X., Vaccine Adjuvants: Mechanisms and Platforms. *Sig. Transduct. Target Ther.*, *8*, 283, 2023, doi: 10.1038/s41392-023-01557-7.
4. Facciolà, A., Visalli, G., Laganà, A., Di Pietro, A., An Overview of Vaccine Adjuvants: Current Evidence and Future Perspectives. *Vaccines*, *10*, 819, 2022, doi: 10.3390/vaccines10050819.
5. Kim, J.Y., Rosenberger, M.G., Rutledge, N.S., Esser-Kahn, A.P., Next-Generation Adjuvants: Applying Engineering Methods to Create and Evaluate Novel Immunological Responses. *Pharmaceutics*, *15*, 1687, 2023, doi: 10.3390/pharmaceutics15061687.
6. He, Y., Xiang, Z., Mobley, H.L.T., Vaxign: The First Web-Based Vaccine Design Program for Reverse Vaccinology and Applications for Vaccine Development. *J. Biomed. Biotechnol.*, *2010*, 1–15, 2010, doi: 10.1155/2010/297505.
7. Xiang, Z., Todd, T., Ku, K.P., Kovacic, B.L., Larson, C.B., Chen, F., Hodges, A.P., Tian, Y., Olenzek, E.A., Zhao, B. *et al.*, VIOLIN: Vaccine Investigation and Online Information Network. *Nucleic Acids Res.*, *36*, D923–D928, 2007, doi: 10.1093/nar/gkm1039.
8. Sayers, S., Ulysse, G., Xiang, Z., He, Y., Vaxjo: A Web-Based Vaccine Adjuvant Database and Its Application for Analysis of Vaccine Adjuvants and Their Uses in Vaccine Development. *J. Biomed. Biotechnol.*, *2012*, 1–13, 2012, doi: 10.1155/2012/831486.
9. Onyango, O.H., Mwenda, C.M., Gitau, G., Muoma, J., Okoth, P., *In-Silico* Analysis of Potent Mosquirix Vaccine Adjuvant Leads. *J. Genet. Eng. Biotechnol.*, *21*, 155, 2023, doi: 10.1186/s43141-023-00590-x.
10. Gogoi, N.R., Bezbaruah, R., Patel, V., Satasia, R., Bhattacharjee, B., Mazumder, B., Regulatory Pathways for Nanocarrier Vaccine, in: *Nanocarrier Vaccines*, V.P. Chavda, and V. Apostolopoulos (Eds.), pp. 465–485, Wiley, Hoboken, New Jersey, U.S., 2024, ISBN 978-1-394-17468-3.
11. Oli, A.N., Obialor, W.O., Ifeanyichukwu, M.O., Odimegwu, D.C., Okoyeh, J.N., Emechebe, G.O., Adejumo, S.A., Ibeanu, G.C., Immunoinformatics and Vaccine Development: An Overview. *ITT*, *9*, 13–30, 2020, doi: 10.2147/ITT.S241064.
12. Woodruff, M.C., Faliti, C.E., Sanz, I., Systems Biology of B Cells in COVID-19. *Semin. Immunol.*, *72*, 101875, 2024, doi: 10.1016/j.smim.2024.101875.
13. Hashemzadeh, P., Ghorbanzadeh, V., Lashgarian, H.E., Kheirandish, F., Dariushnejad, H., Harnessing Bioinformatic Approaches to Design Novel

Multi-Epitope Subunit Vaccine Against Leishmania Infantum. *Int. J. Pept. Res. Ther.*, *26*, 1417–1428, 2020, doi: 10.1007/s10989-019-09949-6.

14. Bezbaruah, R., Chavda, V.P., Nongrang, L., Alom, S., Deka, K., Kalita, T., Ali, F., Bhattacharjee, B., Vora, L., Nanoparticle-Based Delivery Systems for Vaccines. *Vaccines*, *10*, 1946, 2022, doi: 10.3390/vaccines10111946.
15. Bhattacharjee, B., Ikbal, A.M.A., Farooqui, A., Sahu, R.K., Ruhi, S., Syed, A., Miatmoko, A., Khan, D., Khan, J., Superior Possibilities and Upcoming Horizons for Nanoscience in COVID-19: Noteworthy Approach for Effective Diagnostics and Management of SARS-CoV-2 Outbreak. *Chem. Pap.*, *77*, 4107–4130, 2023, doi: 10.1007/s11696-023-02795-3.
16. Raoufi, E., Hemmati, M., Eftekhari, S., Khaksaran, K., Mahmodi, Z., Farajollahi, M.M., Mohsenzadegan, M., Epitope Prediction by Novel Immunoinformatics Approach: A State-of-the-Art Review. *Int. J. Pept. Res. Ther.*, *26*, 1155–1163, 2020, doi: 10.1007/s10989-019-09918-z.
17. Kumar, N., Bajiya, N., Patiyal, S., Raghava, G.P.S., Multi-perspectives and Challenges in Identifying B-cell Epitopes. *Protein Sci.*, *32*, e4785, 2023, doi: 10.1002/pro.4785.
18. Galanis, K.A., Nastou, K.C., Papandreou, N.C., Petichakis, G.N., Pigis, D.G., Iconomidou, V.A., Linear B-Cell Epitope Prediction for *In Silico* Vaccine Design: A Performance Review of Methods Available via Command-Line Interface. *IJMS*, *22*, 3210, 2021, doi: 10.3390/ijms22063210.
19. Nielsen, M., Andreatta, M., Peters, B., Buus, S., Immunoinformatics: Predicting Peptide–MHC Binding. *Annu. Rev. Biomed. Data Sci.*, *3*, 191–215, 2020, doi: 10.1146/annurev-biodatasci-021920-100259.
20. Vielhaben, J., Wenzel, M., Samek, W., Strodthoff, N., USMPep: Universal Sequence Models for Major Histocompatibility Complex Binding Affinity Prediction. *BMC Bioinf.*, *21*, 279, 2020, doi: 10.1186/s12859-020-03631-1.
21. Koşaloğlu-Yalçın, Z., Lee, J., Greenbaum, J., Schoenberger, S.P., Miller, A., Kim, Y.J., Sette, A., Nielsen, M., Peters, B., Combined Assessment of MHC Binding and Antigen Abundance Improves T Cell Epitope Predictions. *iScience*, *25*, 103850, 2022, doi: 10.1016/j.isci.2022.103850.
22. Laimer, J. and Lackner, P., MHCII3D—Robust Structure Based Prediction of MHC II Binding Peptides. *IJMS*, *22*, 12, 2020, doi: 10.3390/ijms22010012.
23. De Ruyck, J., Brysbaert, G., Blossey, R., Lensink, M., Molecular Docking as a Popular Tool in Drug Design, an *in Silico* Travel. *AABC*, *9*, 1–11, 2016, doi: 10.2147/AABC.S105289.
24. Hadni, H. and Elhallaoui, M., 2D and 3D-QSAR, Molecular Docking and ADMET Properties *in Silico* Studies of Azaaurones as Antimalarial Agents. *New J. Chem.*, *44*, 6553–6565, 2020, doi: 10.1039/C9NJ05767F.
25. Langer, T. and Hoffmann, R., Virtual Screening An Effective Tool for Lead Structure Discovery. *CPD*, *7*, 509–527, 2001, doi: 10.2174/1381612013397861.
26. Morris, G.M. and Lim-Wilby, M., Molecular Docking, in: *Molecular Modeling of Proteins, Methods in Molecular Biology*, vol. 443, A. Kukol (Ed.), pp. 365–382, Humana Press, Totowa, NJ, 2008, ISBN 978-1-58829-864-5.

27. Torres, P.H.M., Sodero, A.C.R., Jofily, P., Silva-Jr, F.P., Key Topics in Molecular Docking for Drug Design. *IJMS*, *20*, 4574, 2019, doi: 10.3390/ijms20184574.
28. Meng, X.-Y., Zhang, H.-X., Mezei, M., Cui, M., Molecular Docking: A Powerful Approach for Structure-Based Drug Discovery. *CAD*, *7*, 146–157, 2011, doi: 10.2174/157340911795677602.
29. Muhammed, M.T. and Aki-Yalcin, E., Molecular Docking: Principles, Advances, and Its Applications in Drug Discovery. *LDDD*, *21*, 480–495, 2024, doi: 10.2174/1570180819666220922103109.
30. Desai, S.A., Patel, V.P., Shinde, S.U., Kadam, S.S., *In Silico* Studies for the Bioremediation of Heavy Metals from Contaminated Sites, in: *Microbiome-Assisted Bioremediation*, pp. 139–148, Elsevier, Cambridge, Massachusetts, 2024, ISBN 978-0-443-21911-5.
31. Cáceres, E.L., Tudor, M., Cheng, A.C., Deep Learning Approaches in Predicting ADMET Properties. *Future Med. Chem.*, *12*, 1995–1999, 2020, doi: 10.4155/fmc-2020-0259.
32. Salo-Ahen, O.M.H., Alanko, I., Bhadane, R., Bonvin, A.M.J.J., Honorato, R.V., Hossain, S., Juffer, A.H., Kabedev, A., Lahtela-Kakkonen, M., Larsen, A.S. *et al.*, Molecular Dynamics Simulations in Drug Discovery and Pharmaceutical Development. *Processes*, *9*, 71, 2020, doi: 10.3390/pr9010071.
33. Oren, O., Gersh, B.J., Bhatt, D.L., Artificial Intelligence in Medical Imaging: Switching from Radiographic Pathological Data to Clinically Meaningful Endpoints. *Lancet Digital Health*, *2*, e486–e488, 2020, doi: 10.1016/S2589-7500(20)30160-6.
34. Rubin, D.L., Akdogan, M.U., Altindag, C., Alkim, E., ePAD: An Image Annotation and Analysis Platform for Quantitative Imaging. *Tomography*, *5*, 170–183, 2019, doi: 10.18383/j.tom.2018.00055.
35. Kalpathy-Cramer, J., Zhao, B., Goldgof, D., Gu, Y., Wang, X., Yang, H., Tan, Y., Gillies, R., Napel, S.A., Comparison of Lung Nodule Segmentation Algorithms: Methods and Results from a Multi-Institutional Study. *J. Digit. Imaging*, *29*, 476–487, 2016, doi: 10.1007/s10278-016-9859-z.
36. Echegaray, S., Bakr, S., Rubin, D.L., Napel, S., Quantitative Image Feature Engine (QIFE): An Open-Source, Modular Engine for 3D Quantitative Feature Extraction from Volumetric Medical Images. *J. Digit. Imaging*, *31*, 403–414, 2018, doi: 10.1007/s10278-017-0019-x.
37. Van Griethuysen, J.J.M., Fedorov, A., Parmar, C., Hosny, A., Aucoin, N., Narayan, V., Beets-Tan, R.G.H., Fillion-Robin, J.-C., Pieper, S., Aerts, H.J.W.L., Computational Radiomics System to Decode the Radiographic Phenotype. *Cancer Res.*, *77*, e104–e107, 2017, doi: 10.1158/0008-5472.CAN-17-0339.
38. Aerts, H.J.W.L., Velazquez, E.R., Leijenaar, R.T.H., Parmar, C., Grossmann, P., Carvalho, S., Bussink, J., Monshouwer, R., Haibe-Kains, B., Rietveld, D. *et al.*, Decoding Tumour Phenotype by Noninvasive Imaging Using a Quantitative Radiomics Approach. *Nat. Commun.*, *5*, 4006, 2014, doi: 10.1038/ncomms5006.
39. Kubben, P., Dumontier, M., Dekker, A. (Eds.), *Fundamentals of Clinical Data Science*, Springer International Publishing, Cham, 2019, ISBN 978-3-319-99712-4.

40. Parmar, C., Grossmann, P., Rietveld, D., Rietbergen, M.M., Lambin, P., Aerts, H.J.W.L., Radiomic Machine-Learning Classifiers for Prognostic Biomarkers of Head and Neck Cancer. *Front. Oncol.*, *5*, 272, 2015, doi: 10.3389/fonc.2015.00272.
41. Kar, S. and Leszczynski, J., Exploration of Computational Approaches to Predict the Toxicity of Chemical Mixtures. *Toxics*, *7*, 15, 2019, doi: 10.3390/toxics7010015.
42. Maunz, A., Gütlein, M., Rautenberg, M., Vorgrimmler, D., Gebele, D., Helma, C., Lazar: A Modular Predictive Toxicology Framework. *Front. Pharmacol.*, 9, 4, 38, 2013, doi: 10.3389/fphar.2013.00038.
43. Patlewicz, G., Jeliazkova, N., Safford, R.J., Worth, A.P., Aleksiev, B., An Evaluation of the Implementation of the Cramer Classification Scheme in the Toxtree Software. *SAR QSAR Environ. Res.*, *19*, 495–524, 2008, doi: 10.1080/10629360802083871.
44. Ponting, D.J., Burns, M.J., Foster, R.S., Hemingway, R., Kocks, G., MacMillan, D.S., Shannon-Little, A.L., Tennant, R.E., Tidmarsh, J.R., Yeo, D.J., Use of Lhasa Limited Products for the *In Silico* Prediction of Drug Toxicity, in: *In Silico Methods for Predicting Drug Toxicity, Methods in Molecular Biology*, vol. 2425, E. Benfenati (Ed.), pp. 435–478, Springer US, New York, NY, 2022, ISBN 978-1-07-161959-9.
45. Williams, A.J., Lambert, J.C., Thayer, K., Dorne, J.-L.C.M., Sourcing Data on Chemical Properties and Hazard Data from the US-EPA CompTox Chemicals Dashboard: A Practical Guide for Human Risk Assessment. *Environ. Int.*, *154*, 106566, 2021, doi: 10.1016/j.envint.2021.106566.
46. Zhou, L., Fan, D., Yin, W., Gu, W., Wang, Z., Liu, J., Xu, Y., Shi, L., Liu, M., Ji, G., Comparison of Seven *in Silico* Tools for Evaluating of Daphnia and Fish Acute Toxicity: Case Study on Chinese Priority Controlled Chemicals and New Chemicals. *BMC Bioinf.*, *22*, 151, 2021, doi: 10.1186/s12859-020-03903-w.
47. Shi, W., Guo, J., Bao, T., QSAR Tools for Toxicity Prediction in Risk Assessment—Comparative Analysis, in: *QSAR in Safety Evaluation and Risk Assessment*, pp. 203–218, Elsevier, Cambridge, Massachusetts, 2023, ISBN 978-0-443-15339-6.
48. Li, D. and Wu, M., Pattern Recognition Receptors in Health and Diseases. *Sig. Transduct. Target. Ther.*, *6*, 291, 2021, doi: 10.1038/s41392-021-00687-0.
49. Wilkinson, B.L., Day, S., Malins, L.R., Apostolopoulos, V., Payne, R.J., Self-Adjuvanting Multicomponent Cancer Vaccine Candidates Combining Per-Glycosylated MUC1 Glycopeptides and the Toll-like Receptor 2 Agonist Pam $_3$ CysSer. *Angew. Chem. Int. Ed.*, *50*, 1635–1639, 2011, doi: 10.1002/anie.201006115.
50. Wilkinson, B.L., Day, S., Chapman, R., Perrier, S., Apostolopoulos, V., Payne, R.J., Synthesis and Immunological Evaluation of Self-Assembling and Self-Adjuvanting Tricomponent Glycopeptide Cancer-Vaccine Candidates. *Chem. A Eur. J.*, *18*, 16540–16548, 2012, doi: 10.1002/chem.201202629.

51. Feairheller, S.H., Taylor, M.M., Bailey, D.G., 35S-Sulfide Incorporation during Alkaline Treatment of Keratin and Its Relation to Lanthionine Formation. *Adv. Exp. Med. Biol.*, *86B*, 177–186, 1977, doi: 10.1007/978-1-4757-9113-6_11.
52. Sheng, K., Kalkanidis, M., Pouniotis, D.S., Esparon, S., Tang, C.K., Apostolopoulos, V., Pietersz, G.A., Delivery of Antigen Using a Novel Mannosylated Dendrimer Potentiates Immunogenicity *in Vitro* and *in Vivo*. *Eur. J. Immunol.*, *38*, 424–436, 2008, doi: 10.1002/eji.200737578.
53. Tang, C.K., Lodding, J., Minigo, G., Pouniotis, D.S., Plebanski, M., Scholzen, A., McKenzie, I.F.C., Pietersz, G.A., Apostolopoulos, V., Mannan-mediated Gene Delivery for Cancer Immunotherapy. *Immunology*, *120*, 325–335, 2007, doi: 10.1111/j.1365-2567.2006.02506.x.
54. Yang, J.-X., Tseng, J.-C., Yu, G.-Y., Luo, Y., Huang, C.-Y.F., Hong, Y.-R., Chuang, T.-H., Recent Advances in the Development of Toll-like Receptor Agonist-Based Vaccine Adjuvants for Infectious Diseases. *Pharmaceutics*, *14*, 423, 2022, doi: 10.3390/pharmaceutics14020423.
55. Mahmoodi, S., Amirzakaria, J.Z., Ghasemian, A., *In Silico* Design and Validation of a Novel Multi-Epitope Vaccine Candidate against Structural Proteins of Chikungunya Virus Using Comprehensive Immunoinformatics Analyses. *PLoS One*, *18*, e0285177, 2023, doi: 10.1371/journal.pone.0285177.
56. Mitra, D., Pandey, J., Jain, A., Swaroop, S., *In Silico* Design of Multi-Epitope-Based Peptide Vaccine against SARS-CoV-2 Using Its Spike Protein. *J. Biomol. Struct. Dyn.*, *40*, 5189–5202, 2022, doi: 10.1080/07391102.2020.1869092.
57. Shey, R.A., Ghogomu, S.M., Esoh, K.K., Nebangwa, N.D., Shintouo, C.M., Nongley, N.F., Asa, B.F., Ngale, F.N., Vanhamme, L., Souopgui, J., *In-Silico* Design of a Multi-Epitope Vaccine Candidate against Onchocerciasis and Related Filarial Diseases. *Sci. Rep.*, *9*, 4409, 2019, doi: 10.1038/s41598-019-40833-x.
58. Meza, B., Ascencio, F., Sierra-Beltrán, A.P., Torres, J., Angulo, C.A., Novel Design of a Multi-Antigenic, Multistage and Multi-Epitope Vaccine against Helicobacter Pylori: An *in Silico* Approach. *Infection, Genet. Evol.*, *49*, 309–317, 2017, doi: 10.1016/j.meegid.2017.02.007.
59. Ma, J., Qiu, J., Wang, S., Ji, Q., Xu, D., Wang, H., Wu, Z., Liu, Q.A., Novel Design of Multi-Epitope Vaccine Against Helicobacter Pylori by Immunoinformatics Approach. *Int. J. Pept. Res. Ther.*, *27*, 1027–1042, 2021, doi: 10.1007/s10989-020-10148-x.
60. Yin, Z., Li, M., Niu, C., Yu, M., Xie, X., Haimiti, G., Guo, W., Shi, J., He, Y., Ding, J. *et al.*, Design of Multi-Epitope Vaccine Candidate against Brucella Type IV Secretion System (T4SS). *PLoS One*, *18*, e0286358, 2023, doi: 10.1371/journal.pone.0286358.
61. Department of General Surgery, Ruijin Hospital, Shanghai Institute of Digestive Surgery, Shanghai Key Laboratory for Gastric Neoplasms, Shanghai Jiao Tong University School of Medicine, Shanghai 200025, China and Yu, Y., Repurposing Glucocorticoids as Adjuvant Reagents for Immune Checkpoint

Inhibitors in Solid Cancers. *Cancer Biol. Med.*, *18*, 944–948, 2021, doi: 10.20892/j.issn.2095-3941.2021.0491.

62. Abu Eid, R., Razavi, G.S.E., Mkrtichyan, M., Janik, J., Khleif, S.N., Old-School Chemotherapy in Immunotherapeutic Combination in Cancer, A Low-Cost Drug Repurposed. *Cancer Immunol. Res.*, *4*, 377–382, 2016, doi: 10.1158/2326-6066.CIR-16-0048.
63. Wang, D., Quiros, J., Mahuron, K., Pai, C.-C., Ranzani, V., Young, A., Silveria, S., Harwin, T., Abnousian, A., Pagani, M. *et al.*, Targeting EZH2 Reprograms Intratumoral Regulatory T Cells to Enhance Cancer Immunity. *Cell Rep.*, *23*, 3262–3274, 2018, doi: 10.1016/j.celrep.2018.05.050.
64. Issa, N.T., Stathias, V., Schürer, S., Dakshanamurthy, S., Machine and Deep Learning Approaches for Cancer Drug Repurposing. *Semin. Cancer Biol.*, *68*, 132–142, 2021, doi: 10.1016/j.semcancer.2019.12.011.
65. Apostolopoulos, V., Getting into the Groove: Unusual Features of Peptide Binding to MHC Class I Molecules and Implications in Vaccine Design. *Front. Biosci.*, *6*, d1311, 2001, doi: 10.2741/Apostol.
66. Apostolopoulos, V., Peptide-Based Vaccines for Cancer: Are We Choosing the Right Peptides? *Expert Rev. Vaccines*, *8*, 259–260, 2009, doi: 10.1586/14760584.8.3.259.
67. Apostolopoulos, V., Yu, M., Corper, A.L., Li, W., McKenzie, I.F.C., Teyton, L., Wilson, I.A., Crystal Structure of a Non-Canonical High Affinity Peptide Complexed with MHC Class I: A Novel Use of Alternative Anchors. *J. Mol. Biol.*, *318*, 1307–1316, 2002, doi: 10.1016/S0022-2836(02)00198-5.
68. Lazoura, E., Lodding, J., Farrugia, W., Ramsland, P.A., Stevens, J., Wilson, I.A., Pietersz, G.A., Apostolopoulos, V., Enhanced Major Histocompatibility Complex Class I Binding and Immune Responses through Anchor Modification of the Non-canonical Tumour-associated Mucin 1-8 Peptide. *Immunology*, *119*, 306–316, 2006, doi: 10.1111/j.1365-2567.2006.02434.x.
69. Apostolopoulos, V., Lofthouse, S.A., Popovski, V., Chelvanayagam, G., Sandrin, M.S., McKenzie, F.C., Peptide Mimics of a Tumor Antigen Induce Functional Cytotoxic T Cells. *Nat. Biotechnol.*, *16*, 276–280, 1998, doi: 10.1038/nbt0398-276.
70. Apostolopoulos, V., Osinski, C., Mckenzie, I.F.C., MUC1 Cross-Reactive Galα(l,3)Gal Antibodies in Humans Switch Immune Responses from Cellular to Humoral. *Nat. Med.*, *4*, 315–320, 1998, doi: 10.1038/nm0398-315.
71. Matsoukas, J., Apostolopoulos, V., Kalbacher, H., Papini, A.-M., Tselios, T., Chatzantoni, K., Biagioli, T., Lolli, F., Deraos, S., Papathanassopoulos, P. *et al.*, Design And Synthesis of a Novel Potent Myelin Basic Protein Epitope 87–99 Cyclic Analogue: Enhanced Stability and Biological Properties of Mimics Render Them a Potentially New Class of Immunomodulators. *J. Med. Chem.*, *48*, 1470–1480, 2005, doi: 10.1021/jm040849g.
72. Tang, C.-K., Katsara, M., Apostolopoulos, V., Strategies Used for MUC1 Immunotherapy: Human Clinical Studies. *Expert Rev. Vaccines*, *7*, 963–975, 2008, doi: 10.1586/14760584.7.7.963.

3

Adjuvant and Immunity

Himel Mondal[1], Shaikat Mondal[2], Bairong Shen[3] and Rajeev K. Singla[3,4*]

[1]*Department of Physiology, All India Institute of Medical Sciences, Deoghar, Jharkhand, India*
[2]*Department of Physiology, Raiganj Government Medical College and Hospital, West Bengal, India*
[3]*Department of Pharmacy and Institutes for Systems Genetics, Center for High Altitude Medicine, Frontiers Science Center for Disease-Related Molecular Network, West China Hospital, Sichuan University, Chengdu, Sichuan, China*
[4]*School of Pharmaceutical Sciences, Lovely Professional University, Phagwara, Punjab, India*

Abstract

Adjuvants play a crucial role in enhancing the immune response triggered by vaccines, thereby improving vaccine efficacy and durability of protection. This chapter explores the intricate relationship between vaccine adjuvants and the immune system, providing a comprehensive understanding of how adjuvants enhance immune responses triggered by vaccines. The chapter discusses the impact of adjuvants on humoral and cellular immunity, highlighting their role in enhancing antibody production, generating memory T cells, and activating cytotoxic T lymphocytes. Additionally, it explores the influence of adjuvants on innate immunity and their potential to promote mucosal immune responses. The chapter addresses adjuvants on immunity and their significance in improving vaccine efficacy and public health. It emphasizes the importance of adjuvants in vaccine durability and their potential to provide long-lasting immunity.

Keywords: Antigen presentation, adjuvants, vaccine, immunology, humoral immunity, cellular immunity

**Corresponding author*: rajeevsingla26@gmail.com

Vivek P. Chavda and Vasso Apostolopoulos (eds.) Emerging Pathways of Vaccine Adjuvants: A Nonspecific Stimulant of the Immune System, (39–72) © 2025 Scrivener Publishing LLC

3.1 Introduction

Adjuvants play a crucial role in enhancing the effectiveness of vaccines by stimulating the immune system's response to antigens present in vaccines. These specialized compounds are included in vaccine formulations to amplify the immune response, leading to more robust and longer-lasting protection against infections. Adjuvants mimic pathogens' signals, promoting a more robust immune system activation. By engaging specific receptors on immune cells, adjuvants help increase the production of antibodies, activate T cells, and generate memory responses. This heightened immune reaction is significant when dealing with weak or poorly immunogenic antigens, allowing for a more potent defense mechanism to be established [1].

The interaction between adjuvants and the immune system is complex and dynamic. Adjuvants enhance the immediate response to the antigen and influence the nature and quality of the formed immune memory. They guide the immune system toward a specific response, whether a predominantly antibody-based response (humoral immunity) or a cellular immune response. Moreover, adjuvants can also modulate the balance between different types of immune cells, such as helper T cells, regulatory T cells, and various subsets of immune cells, ensuring a well-coordinated and controlled immune reaction [2].

However, the interaction between adjuvants and the immune system must be carefully balanced. While adjuvants boost immune responses, an excessive or inappropriate immune activation could lead to adverse effects [3]. This underscores the importance of thorough testing and safety assessment of adjuvants before they are approved in vaccines. Researchers continue investigating and developing novel adjuvants that can efficiently stimulate immune responses while minimizing potential risks. Understanding the intricate interplay between adjuvants and the immune system is fundamental to designing vaccines that safeguard public health against infectious diseases [4].

This chapter will explore the multifaceted role of adjuvants in modulating immunity in the following section:

Immune Response to Vaccines: This section discusses the process of natural immunity and immunity produced by vaccines and how adjuvants play a pivotal role in modulating and enhancing these responses. By elucidating the mechanisms by which adjuvants interact with different facets of the immune system, we gain insights into how vaccines can be tailored to elicit robust and lasting protection.

Mechanisms of Adjuvants in Modulating Immunity: The efficacy of vaccines is often influenced by the immune system's ability to recognize, remember, and combat specific pathogens. Adjuvants, as immunostimulatory agents, intricately shape these processes. From engaging critical receptors on immune cells to facilitating antigen presentation, the multifaceted interactions between adjuvants and immunity are dissected in this section to provide a comprehensive understanding of their immunomodulatory role.

Immunogenicity According to the Types of Adjuvants: Different types of adjuvants exert distinct effects on the immune system, leading to varying degrees of immunogenicity. By delineating how different adjuvant classes stimulate adaptive immunity, we gain insights into tailoring vaccine formulations for specific pathogens and target populations.

Adjuvants and Humoral Immunity: Humoral immunity, characterized by the production of antibodies, is a fundamental defense mechanism against extracellular pathogens. By elucidating the synergy between adjuvants and B cells, we uncover strategies to design vaccines that induce potent antibody-mediated protection.

Adjuvants and Cellular Immunity: Cellular immunity by T cells plays a pivotal role in combating intracellular pathogens and providing immunity. By dissecting how adjuvants influence T-cell activation, differentiation, and memory formation, we understand how a robust cellular immunity for a comprehensive defense against infections is developed.

Adjuvants and Innate Immunity: The rapid and generalized responses of the innate immune system serve as the initial line of defense against pathogens. This section discusses how adjuvants activate innate immune receptors, triggering cascades of events that shape subsequent adaptive responses. By exploring the crosstalk between adjuvants and innate immunity, the foundations of early immune activation pave the way for long-lasting protection.

Adjuvants and Mucosal Immunity: Mucosal surfaces are primary entry points for many pathogens, necessitating the development of vaccines that elicit robust mucosal immunity. This segment examines how adjuvants stimulate mucosal immune responses and highlight their role in enhancing mucosal antibody secretion and providing defense against mucosal infections.

Adjuvants and Vaccine Efficacy in Specific Populations: Vaccine responses can vary among different populations, including infants, elderly individuals, and immunocompromised individuals. This segment discusses how adjuvants can enhance vaccine efficacy in these specific populations.

The remarkable success of vaccines in preventing infectious diseases is anchored in the intricate relationship between adjuvants and the immune system. This chapter explores the multifaceted role of adjuvants in modulating immunity and their importance in vaccine design and development. By influencing innate, cell-mediated, and humoral immunity, adjuvants collectively shape immune responses against antigens present in vaccines.

3.2 Immune Response to Vaccines

The immune response to pathogens and vaccines involves complex interactions between the immune system and the foreign antigens introduced into the body. This segment discusses how the immune system responds to pathogens (disease-causing agents) and vaccines (introduced antigens meant to mimic pathogens). The knowledge would help us understand the body's response to antigens, and how adjuvants alter this mechanism will be discussed later.

3.2.1 Immune Response to Pathogens

The immune system's response involves the engagement of immune cells, such as macrophages and dendritic cells (DCs), which employ the process of phagocytosis to engulf pathogens. This intricate mechanism allows immune cells to internalize pathogens for subsequent processing. Within the confines of phagocytic cells, pathogens undergo a process of enzymatic degradation, breaking these invaders into smaller fragments. This pivotal step neutralizes pathogens and generates antigenic fragments that form the basis for antigen presentation, a cornerstone in initiating adaptive immune responses.

There is an interaction between the pathogen-associated molecular patterns (PAMPs) of pathogens and pattern recognition receptors (PRRs) situated on immune cells. This interaction initiates a molecular alarm, promptly alerting the immune system to an ongoing infection. This initial interaction serves as a fundamental trigger, initiating a cascade of immune reactions that mobilize immune cells, release signaling molecules, and activate intracellular pathways [5]. These provide a coordinated defense against the infiltrating pathogen.

Antigen-presenting cells (APCs), including DCs, play a pivotal role in the immune response by processing antigens derived from pathogens.

These antigens are showcased on the cell surface by utilizing major histocompatibility complex (MHC) molecules. This presentation holds the key to activating other immune cells, particularly T cells. The interaction between the presented antigen and T-cell receptors triggers a cascade of immune responses, culminating in the coordination of adaptive immune reactions that target and neutralize the pathogenic threat [6].

Helper T cells (CD4+) are central to T-cell and B-cell immunity. Following activation, these helper T cells unleash a cascade of immune activities by releasing a spectrum of cytokines. These cytokines wield the power to steer diverse immune responses, such as the mobilization of immune cells, the modulation of inflammation, and the augmentation of antibody production. In addition, helper T cells extend their assistance beyond their realm, providing critical support to other immune components [7]. They lend a helping hand to B cells, fostering the production of antibodies and enhancing cytotoxic T cells' functionality.

B cells possess a twofold mechanism for pathogen recognition. They can directly identify pathogens by binding their surface antibodies to pathogen-specific antigens. Alternatively, B cells receive assistance from helper T cells, which recognize the antigen–MHC complex on APCs and release cytokines that help B-cell activation. This activation initiates a transformative journey, leading B cells to differentiate into plasma cells. These specialized plasma cells serve as antibody factories, producing antibodies that are exquisitely tailored to the encountered pathogen. These antibodies neutralize the pathogen by impeding its infective capacity and perform a vital tagging function. The tagged pathogens are then marked for destruction by other immune cells [8].

Cytotoxic T cells (CD8+) are integral components of cellular immunity. It possesses a remarkable ability to detect and eliminate infected cells. This recognition hinges on their capacity to identify pathogen-derived antigens displayed on the surface of infected cells. This antigen presentation occurs through the interaction between antigen–MHC complexes and the T-cell receptor on cytotoxic T cells. Once the specific recognition event occurs, cytotoxic T cells neutralize the infected cells. These include the release of cytotoxic granules containing perforin and granzymes, which induce apoptosis in the infected cells. This proactive response is strategically designed to halt the replication of pathogens within the infected cells, thereby curbing the further dissemination of the infection.

Following the successful clearance of an infection, memory T cells and memory B cells are established. These specialized cells retain a crucial role in the body's immunity by embodying a form of immunological memory. Their formation is a testament to the immune system's adaptive capacity, as they "remember" the encountered pathogen. This memory imprint translates into a profound advantage upon subsequent exposures to the same pathogen. During these subsequent encounters, memory T and B cells show a more potent response than the initial response during the primary infection. The capacity of these memory cells to expedite immune reactions encapsulates the dynamic interplay between innate and adaptive immunity, providing the immune system with a vigilant defense mechanism against recurring threats [9].

3.2.2 Immune Response to Vaccines

Vaccines are an artificial infection that also triggers the body's immune system. They encompass attenuated, inactivated, or fragmented forms of pathogens and synthetic antigens. Vaccines introduce these antigens to the immune system upon administration, offering a controlled encounter that does not provoke the full-fledged disease. This exposure triggers the immune system's recognition and subsequent response to the antigens, leading to the generation of immune memory. This memory serves as a blueprint for mounting rapid and potent immune reactions if the individual encounters the actual pathogen in the future. Hence, vaccines are intentionally introduced to protect the body from infection.

Dendritic cells emerge as instrumental players in the processing of vaccine antigens. The antigens derived from vaccines containing weakened, inactivated, or fragmented pathogens are captured and internalized by DCs. Subsequently, DCs undergo a process of antigen processing, wherein the antigens are broken down into smaller fragments. These processed antigens are then showcased on the surface of DCs through MHC molecules. This recognition triggers a cascade of events that culminate in the activation of T cells. This pivotal interaction between DCs, vaccine antigens, and T cells encapsulates the essence of the adaptive immune response [10].

Helper T cells exhibit a critical role in recognizing antigens presented by APCs, instigating a multifaceted array of immune responses. By releasing a repertoire of cytokines, these T cells modulate the activation of B cells, facilitating antibody production and enhancing the proliferation and activation of cytotoxic T cells.

Upon recognizing antigens, B cells undergo differentiation, maturing into plasma cells that specialize in producing antibodies targeting the encountered antigens. These antibodies are a potent force against future pathogenic encounters, as they are poised to neutralize pathogens that carry the same antigens as those found in the vaccine.

Like their role in pathogen encounters, the immune system helps form memory T and B cells following vaccination, thus ensuring a proficient defense strategy. These specialized cells embody immunological memory, permitting swift and efficient responses upon subsequent exposures to the same antigens. The memory T cells and memory B cells, having "remembered" the encountered vaccine antigens and expedite the immune reaction [11].

Certain vaccines necessitate multiple doses, often called booster shots, to fortify the immune response. These additional doses are pivotal in enhancing the potency and durability of the generated immune memory. The immune system's response to initial vaccination may wane over time; hence, booster shots act as timely reminders, reinvigorating the immune cells' memory and reinforcing the defense mechanisms. By strategically administering booster shots, the immune system can mount more substantial and enduring immune responses, thus elevating vaccine-induced immunity's overall efficacy and sustainability [12].

In both cases (pathogen and vaccine), understanding the interaction between antigens and the immune system and the roles of different immune cells and molecules is critical for vaccine development, immunotherapy, and controlling infectious diseases.

3.3 Mechanisms of Adjuvants in Modulating Immunity

This chapter describes the interaction between adjuvants and the immune system. We explore the multifaceted ways in which adjuvants influence immune reactions. Below is the description of the mechanism as mentioned above.

3.3.1 Sustained Release of Antigen

The concept of depot formation at the injection site represents one of the earliest and most universally acknowledged mechanisms through which adjuvants exert their effects. This phenomenon involves the entrapment of antigens, gradually releasing them at the injection site, thereby maintaining continual immune system stimulation and promoting robust antibody production. Historically, the depot effect was widely considered a fundamental mechanism underlying the activity of various adjuvants. For instance, the significance of depot formation in the adjuvant activity of alum was initially proposed. Several other adjuvants, including water-in-oil emulsions such as complete Freund's adjuvant (CFA) and biodegradable micro- and nanoparticles (NPs), have also been demonstrated to rely on the depot effect to induce sustained high levels of antibody production. A combination of monophosphoryl lipid A (MPL) and alum showed optimal immune response only when the antigen was co-localized with alum, stabilizing MPL and the antigen and providing a depot effect [13, 14].

Contrary to prior beliefs, there is no definitive evidence to conclusively establish the significant role of the depot effect in adjuvant activity. Notably, antigen adsorption onto alum was not a requisite for alum's adjuvant activity. Recent findings revealed that removing the injection site shortly after antigen and alum administration did not affect humoral or cell-mediated immunity [15].

3.3.2 Upregulation of Cytokines and Chemokines

Particulate adjuvants create local inflammation that draws immune cells, fostering a pro-inflammatory environment for recruitment. Genome-wide analysis revealed that alum, MF59, and CpG-ODN modulate genes associated with cytokines, chemokines, innate immune receptors, interferon-induced genes, and adhesion molecules, termed "adjuvant core response genes". MF59 demonstrated strong modulation of these genes compared to alum and CpG-ODN. Chemokines, crucial for immune cell migration, were upregulated by adjuvants. MF59 particularly elevated CCR2 expression, involved in monocyte infiltration. AS03, another adjuvant, co-localized with antigen to trigger cytokines, chemokines, and chemokine receptor expression at the injection site. Alum and MF59 induced immune cell infiltration, accompanied by chemo-attractant production, supporting their role in immune cell recruitment.

Alum's induction of Th2-type responses and B-cell differentiation enhances antibody production. *In vivo* studies demonstrated alum-induced interleukin-4 (IL-4)–producing Gr1+ cell priming, vital for antigen-specific B-cell proliferation and optimal antibody production. Alum and MF59 activate the complement cascade, creating inflammation and cell recruitment. MF59 also triggers neutrophil, monocyte, eosinophil, and macrophage recruitment, facilitating antigen and adjuvant uptake by recruited cells [16–18].

Adjuvants induce immune cell recruitment, with some cells transporting antigens to lymph nodes for immune response initiation. However, the relationship between recruited cells and immune response induction is intricate. Depletion studies suggest redundancy in the role of recruited cells in adaptive responses, often due to compensatory mechanisms. Detailed investigations are needed to understand the connections between recruited immune cells and adjuvant activity.

3.3.3 Cellular Recruitment at the Site of Injection

Adjuvants create a localized inflammatory environment at the injection site, triggering the influx of various immune cells. This helps recruit immune cell types such as neutrophils, monocytes, macrophages, DCs, and eosinophils.

The presence of these immune cells at the injection site serves multiple purposes. First, it enables rapid interactions between APCs and antigens, facilitating antigen uptake and processing. Second, the inflammatory environment promotes the activation and maturation of APCs, leading to enhanced antigen presentation to T cells. Third, the proximity of immune cells to antigens ensures efficient coordination between innate and adaptive immune responses.

Cellular recruitment is often mediated by chemokines, small signaling molecules that guide immune cells to specific sites. Adjuvants induce the release of chemokines at the injection site, attracting immune cells from the bloodstream and surrounding tissues. This recruitment enhances the chances of immune cells encountering antigens, initiating immune responses, ultimately leading to the generation of memory cells for long-term protection [19].

By creating a conducive microenvironment at the injection site, adjuvants ensure effective interactions between immune cells and antigens, setting the stage for developing a robust and enduring immune defense against potential pathogens.

3.3.4 Increased Antigen Uptake and Presentation

Adjuvants help APCs, such as DCs, macrophages, and B cells, absorb antigens. This increased absorption speeds antigen processing into smaller peptides displayed on MHC molecules. This presentation is critical for activating T cells, essential in both humoral and cellular immune responses.

Adjuvants increase antigen uptake through multiple means. Some adjuvants form complexes with antigens, promoting their efficient internalization by APCs. Others induce cellular activation and inflammation at the injection site, attracting immune cells and increasing the likelihood of encountering antigens. Moreover, adjuvants can modify the physical characteristics of antigens, rendering them more accessible and recognizable to APCs [20].

By enhancing antigen presentation, adjuvants enable efficient activation of helper T cells (CD4+) and cytotoxic T cells (CD8+), leading to a broader and more robust immune response. This mechanism is especially vital for pathogens that can evade immune detection or antigens that are inherently weak stimulators of immune responses.

3.3.5 Activation and Maturation of APCs

Adjuvants are essential in boosting the maturation of APCs like DCs from an immature to a mature, highly efficient condition. This maturation phase is required to initiate and direct adaptive immune responses.

Adjuvants increase the expression of co-stimulatory molecules on the surface of APCs, such as CD80 and CD86. These chemicals are required to improve contact between APCs and T cells, resulting in the successful activation of T cells. Adjuvants also stimulate APCs to produce pro-inflammatory cytokines, producing a favorable milieu for immune cell communication and response coordination.

Adjuvants improve antigen presentation by allowing antigens to be internalized, degraded into smaller pieces, and loaded onto MHC molecules. Mature APCs can then present these antigen–MHC complexes to T cells, boosting helper (CD4+) and cytotoxic (CD8+) T cells [21].

This mechanism is essential for activating adaptive immune responses, as matured APCs efficiently prime T cells and guide them toward their respective roles in humoral and cellular immunity. The concerted actions of adjuvants in activating and maturing APCs amplify the immune response, promoting the production of antibodies, memory cells, and cytotoxic T lymphocytes (CTLs) that provide potent and enduring protection against invading pathogens.

3.3.6 Activation of Inflammasomes

Inflammasomes are multi-protein complexes within cells that are critical in initiating inflammatory responses. Adjuvants can trigger the activation of these complexes, leading to the release of pro-inflammatory cytokines like IL-1β and IL-18.

Adjuvants activate inflammasomes by activating PRRs in immune cells. These PRRs identify PAMPs or danger-associated molecular patterns, causing inflammasome complexes to assemble. The activated inflammasomes then cleave pro–IL-1 and pro–IL-18 into mature versions, ultimately released to stimulate strong immune responses.

Inflammasome activation is crucial for a variety of reasons. First, it boosts the immune response by boosting the production of pro-inflammatory cytokines, which drive immune cell recruitment and activation to the site of infection or vaccination. Second, inflammasomes help in antigen presentation by increasing antigen cross-presentation to cytotoxic T cells. Third, they influence the adaptive immune response by encouraging the development of T helper cell subsets [22].

In conclusion, the interactions between adjuvants and the immune system constitute a pivotal cornerstone in vaccine development. Recent advances have illuminated diverse mechanisms by which adjuvants augment immune responses, including sustained antigen release, cytokine upregulation, cellular recruitment, enhanced antigen uptake and presentation, activation of inflammasomes, and more. Understanding these dynamics empowers scientists to tailor vaccines for specific populations, address immunological challenges, and design novel strategies to combat infectious diseases.

3.4 Immunogenicity According to the Types of Adjuvants

This segment discusses various available adjuvants and their immunogenicity. Adjuvants used for human and veterinary vaccines are included in the discussion [23, 24].

3.4.1 Minerals

Aluminum salts, notably alum, are vital components in many approved human vaccines targeting diseases like human papillomavirus (HPV), hepatitis A virus (HAV), hepatitis B virus (HBV) and others. They even have been investigated for potential use in the formulation of the COVID-19 vaccine. Recent research demonstrated that combining alum with proteins like S or RBD increased IgG1 titers and antibody affinity, leading to long-lasting memory B-cell responses. Adjuvant selection hinges on numerous factors, including disease type, vaccination approach, antigen properties, required immune response, and target population age. A wrong choice of adjuvant can compromise vaccine effectiveness, underscoring the need for careful antigen-adjuvant pairing to optimize efficacy.

Apart from aluminum salts, calcium phosphate is a prominent mineral salt adjuvant, particularly in diphtheria, tetanus, and pertussis vaccines. It prompts IgE and IgG1 responses, influencing the immune pathway toward Th1. Zinc oxide, another biocompatible adjuvant, stimulates a Th2 immune response similar to alum. Cobalt oxide, meanwhile, alters antibody production ratios to favor IgG2α while reducing IgE, mitigating allergic reactions and lymphatic inflammation risks. Other metal salts like iron oxides and tin have garnered less attention as adjuvants. Emerging trends involve exploring the adjuvant potential of gold and silver NPs [25].

3.4.2 Emulsions

Oil-in-water (O/W), water-in-oil (W/O), and a complex of W/O emulsions are all oils in aqueous systems. The W/O emulsions contain complete Freund's adjuvant (CFA) and incomplete Freund's adjuvant (IFA) adjuvants, essential in veterinary vaccinations against foot-and-mouth disease (FMD), equine influenza virus, plague, rabies, and other diseases. Notably, W/O emulsions efficiently reduce antigen dosages in vaccine manufacture, as demonstrated by the protection of broilers against Newcastle disease using mineral oil-based adjuvants even with a lower vaccination dose.

With advancements in CFA and IFA formulations, alternatives such as Montanide VG 51 ISA, Montanide VG 720 ISA, and Adjuvant 65 have been found. Like IFA, W/O emulsions, such as Montanide VG 720 ISA and 51, have therapeutic use in cancer, malaria, AIDS, and autoimmune illnesses. Adjuvant 65, mainly composed of peanut oil, provides dose-reduction advantages while raising concerns about peanut allergies. Widely employed in veterinary medicine, the MF59 adjuvant, an O/W emulsion, harnesses squalene oil alongside non-ionic surfactants (80 Tween and 85 Span), fostering robust immunogenicity and safety in human applications. It fosters an immune-conducive milieu, augmenting specific immune responses to antigens, potentially acting as an antigen delivery system, and enhancing phagocytosis and antigen uptake by APCs. Unlike alum, MF59 showcases heightened immune cell activation, excellent vaccine antigen utilization at the injection site, and elevated APC presence in draining lymph nodes. AS03, another squalene-based O/W emulsion, incorporates alpha-tocopherol to invigorate immune responses. This blend fosters antigen-specific memory cells within draining lymph nodes and evokes combined Th2/Th1 responses. MF59 and AS03 have earned approval for the H1N1 influenza vaccine, heralding a promising era of emulsion adjuvant utilization in enhancing immune responses [26, 27].

3.4.3 Polymers

They are promising for preventing inflammatory diseases and cancers and are even being explored for autoimmune disease research. The performance of these adjuvants depends on factors like solubility, molecular weight, branching, and composition. These polymers may be coupled with live viral and DNA vaccines, providing benefits over standard adjuvants such as biocompatibility, compatibility with live vaccines based on viral or bacterial vectors, biodegradability, safety, and simplicity of manufacture. Over time, natural biodegradable polymers derived from plants and bacteria and synthetic polymers have been employed to improve antigen delivery and reduce vaccination dosages. Fructan, glucan, mannan, and chitosan are natural polymers, whereas polyphosphates, polyelectrolytes, and polyanhydrides are manufactured emulsion polymers. Some carbohydrates also have adjuvant effects because of their immune system signaling responsibilities.

Polymers impact immunity by triggering several mechanisms that enhance the immune response. They improve the presentation of antigens to immune cells by encapsulating and delivering them to APCs. Some polymers induce localized inflammation, attracting immune cells and promoting efficient antigen uptake. Additionally, specific polymers directly activate immune cells like DCs, facilitating antigen presentation. Co-stimulatory signals provided by polymers mimic natural infections, prompting robust immune activation. These polymers can also bias the immune response toward cellular or humoral immunity. They enhance the cross-presentation of antigens, create depots for sustained release, and optimize antigen uptake, collectively contributing to a potent and tailored immune response [28].

3.4.4 Saponins

Saponins, specifically tripeptide glycosides, are plant-extracted adjuvants known for their immune-stimulating properties. Derived from plants like the Gualaya Saponaria tree, these compounds have been used in livestock vaccines, such as those for equine influenza and canine parvovirus. Saponins, like Quil A, can induce cytokine production and enhance immune responses. Quil A, a saponin isolated from QS-21 extract, is notable for its ability to stimulate cellular responses, including CTLs, cytokines like interleukin and interferon, and IgG2a antibodies. ISCOMs (immune stimulating complexes) are cage-like structures containing cholesterol, Quil A, antigens, and phospholipids. These structures, about 30–40 nm in diameter, have provoked a diverse range of cellular and humoral immune responses. ISCOMs can encapsulate hydrophobic or amphipathic antigens, modulate the immune system, and efficiently deliver antigens to B cells, thus increasing antigen uptake. They are used in vaccines for various diseases in cattle and cats. However, it is worth noting that there have been instances of toxic effects associated with ISCOMs in mice [29].

3.4.5 Complement Components and Cytokines

The constituents of the mammalian complement system exhibit promising adjuvant qualities in stimulating antibody responses. Fragments of these proteins bind to foreign antigens, triggering interactions with antibody receptors and immune cells. Notably, the C3d component has shown significant potential. Research indicates that attaching three C3d molecules to an antigen can amplify its immunogenicity by up to 1,000 times, suggesting its valuable role as an effective and distinct adjuvant. Various antigens, including influenza virus hemagglutinin, anti-idiotype antibodies, and capsular polysaccharides, have been successfully enhanced through modifications involving C3d.

Cytokine proteins and their genes have proven to be effective adjuvants in vaccines. IL-2, IL-1, interferon (INF), granulocyte-macrophage colony-stimulating factors (CSF 3-GM), and IL-12 are a few examples. INF- stimulates cellular immune responses via various methods, whereas CSF 3-GM promotes the main immune response by stimulating antigen-supplying cells. When mixed with different adjuvants, recombinant IL-1 and IL-2 have shown potential in cow and sheep vaccinations. Because of its ability to alter immune responses toward type 1, IL-12 is garnering interest. It is utilized in feline immunodeficiency virus vaccinations as an adjuvant. Cytokines are utilized to create vaccines against viral infections in poultry, cattle, and fish. IL-18, for example, is employed in the Newcastle disease vaccine, whereas IL-7 is used in a DNA vaccine against the infectious disease Bursa of Fabricius. Bovine IL-18 boosts cellular and antibody responses in a DNA vaccination for foot-and-mouth disease in cattle. Furthermore, a salmon anemia virus protein and fish type I interferon DNA vaccination generated greater antibody levels and immune cell responses in Atlantic salmon than a cytokine-free DNA vaccine [30].

3.4.6 Bacterial Components

The initial recombinant vaccine expression systems were established using *E. coli* bacteria, offering the advantage of producing substantial quantities of well-defined proteins. However, due to differences in processing mechanisms between prokaryotic and eukaryotic cells, expressed proteins in this system can often be misfolded. Vaccine proteins' signal sequences, glycosylation sites, and disulfide bonds can cause bacteria toxicity, insolubility, or fast disintegration. The functioning of these proteins can be modified by activating the host defense system Toll-like receptor (TLR)–like receptors in response to danger signals. Propionibacterium avidum KP-40–containing heat-killed and lyophilized products can boost antibody responses to co-distributed antigens. Muramyl dipeptide (MDP) is an active molecule in Mycobacterium peptidoglycans with immunomodulatory effects. MDP produces less harmful byproducts, despite potential adverse effects such as fever, arthritis, and iris inflammation. Hydrophilic chemicals, such as threonyl-MDP, enhance type 2 immune responses and are frequently utilized in vaccines for illnesses such as FeLV. Conversely, hydrophobic derivatives are integrated into compounds like liposomes or ethanolamines in water or oil emulsions, designed to provoke type 1 reactions and cellular immunity [31].

3.5 Adjuvants and Humoral Immunity

Adjuvants play a significant role in enhancing humoral immunity, which involves the production of antibodies by B cells. Here is how adjuvants influence humoral immune responses [32].

3.5.1 B-Cell Activation and Antibody Production

Adjuvants help in B-cell activation and the subsequent production of antibodies within the immune response. By aiding in the presentation of antigens, these compounds heighten the efficiency of antigen uptake and processing by APCs. Furthermore, adjuvants supply co-stimulatory signals that activate B cells, a process crucial for their differentiation and proliferation. Additionally, adjuvants stimulate the release of essential cytokines, such as IL-6 and IL-12, further supporting B-cell activation and antibody class switching. By encouraging the formation and maintenance of germinal centers, specialized regions fostering B-cell interactions, and affinity maturation, adjuvants contribute to developing diverse, high-affinity antibodies. Moreover, these compounds influence the process of antibody isotype switching, guiding B cells toward producing specific types of antibodies, such as immunoglobulin G (IgG), essential for robust immunity [33].

3.5.2 Enhanced Germinal Center Formation

Germinal centers are sites where B cells undergo somatic hypermutation and selection processes. Adjuvants enhance these processes by creating an environment that encourages the interaction of B cells, T cells, and APCs. This interaction within germinal centers enables the generation of a diverse repertoire of B cells, each producing antibodies with varying degrees of affinity for the antigen. Adjuvants thus significantly contribute to the formation and maintenance of germinal centers, ultimately leading to a more effective and specific immune response.

3.5.3 Isotype Switching

Adjuvants are found to guide B cells toward producing specific classes of antibodies, known as isotypes. Different antibody isotypes have unique functions in the immune response. For instance, some adjuvants can enhance the production of IgG antibodies. IgG antibodies are essential for neutralizing pathogens by binding to them and preventing their interaction with host cells. Additionally, IgG antibodies facilitate opsonization, a process in which pathogens are coated with antibodies, making them more recognizable to phagocytic cells for engulfment and destruction. Adjuvants contribute to shaping the immune response by influencing antibody class switching, which is crucial for tailoring the immune response to combat diverse types of pathogens effectively [34].

3.5.4 Long-Lasting Antibody Responses

Adjuvants play a critical role in establishing long-lasting immunity by fostering the development of memory B cells. These memory B cells can recognize specific antigens even after the initial immune response has subsided. Memory B cells can quickly differentiate into antibody-producing plasma cells when the body reencounters the same antigen, leading to a rapid and robust immune response. This process allows the immune system to mount a swifter defense upon re-exposure to the pathogen, effectively preventing infection or mitigating its severity. Adjuvants contribute to the creation of memory B cells by enhancing the activation and differentiation of B cells during the primary immune response, which lays the foundation for enduring immunity.

3.5.5 Antibody Affinity Maturation

Affinity maturation is a fundamental mechanism that refines the specificity and effectiveness of antibodies, and adjuvants help antibody affinity maturation. Within germinal centers, B cells undergo a process of somatic hypermutation, during which their antibody genes undergo random mutations. Adjuvants contribute to this process by fostering an environment that supports B-cell survival and proliferation with higher-affinity antibodies. Through subsequent rounds of selection and competition, B cells carrying antibodies with improved affinity for the antigen are favored, whereas those with lower affinity are gradually outcompeted. As a result of affinity maturation, the immune system produces antibodies better suited to target and neutralize the specific antigen, enhancing the overall effectiveness of the immune response. Adjuvants play a significant role in this critical process of antibody optimization by promoting the formation and maintenance of germinal centers.

3.5.6 Enhanced IgG Subclass Responses

Adjuvants help to shape the distribution of different IgG subclasses produced in response to vaccination. Different subclasses of IgG antibodies possess distinct effector functions, making them particularly suited for various immune activities. For instance, some adjuvants can bias the immune response toward producing specific IgG subclasses highly effective at opsonization. Other subclasses may be better at activating the complement system. Adjuvants play a role in directing the immune response toward producing the most suitable subclass or combination of subclasses based on the nature of the pathogen and the desired immune outcome. This capacity to modulate the balance of IgG subclasses further underscores the significance of adjuvants in optimizing the immune response to provide the best protection against specific pathogens [35].

3.5.7 Increased Antibody Titers

Adjuvants help form a higher antibody concentration produced in response to vaccination. This increased antibody level directly results from the adjuvant's ability to enhance the activation and differentiation of B cells, leading to a more outstanding production of antibodies specific to the targeted antigen. Elevated antibody titers significantly strengthen the immune response by ensuring a robust and effective defense against the targeted pathogen.

Adjuvants enhance various aspects of the immune response, including B-cell activation, antibody production, germinal center formation, and memory cell generation. By positively influencing these critical processes, adjuvants contribute significantly to creating vaccines that are more potent and capable of providing enhanced protection against infectious diseases.

3.6 Adjuvants and Cellular Immunity

Adjuvants also play a crucial role in enhancing cellular immunity, activating T cells and their functions in immune responses. While much attention has traditionally been focused on the role of adjuvants in enhancing humoral immunity, recent research has shed light on their crucial contribution to cellular immunity. Cellular immunity, also known as cell-mediated immunity, is one arm of the immune response that activates various immune cells, notably T lymphocytes (T cells). Historically, adjuvants were associated primarily with enhancing antibody responses, but contemporary research has revealed their multifaceted role in promoting cellular immunity [36]. In this section, we discuss how adjuvants influence cellular immune responses.

3.6.1 Activation of Antigen-Presenting Cells (APCs)

Adjuvants can shape the activation and differentiation of T cells, leading to the development of robust and durable cellular immune responses. Mechanisms through which adjuvants stimulate cellular immunity include enhancing the maturation and activation of APCs, such as DCs. Activated APCs effectively process and present antigens to T cells, including cytotoxic T cells (CD8+ T cells) and helper T cells (CD4+ T cells). This process is crucial for priming and activating T cells, essential for targeting infected cells and coordinating immune responses [37]. Adjuvants also influence cytokine production, memory formation, and cross-presentation, all contributing to potent cellular immunity. Examples of adjuvants that enhance cellular immunity include CpG oligodeoxynucleotides, MF59, monophosphoryl lipid A (MPLA), and AS01/AS02 adjuvant systems.

3.6.2 Cytokine Production and T-Cell Differentiation

Adjuvants increase cytokine production by APCs, effectively steering the differentiation of CD4+ T cells into diverse subsets, including Th1, Th2, Th17, and regulatory T cells (Tregs). This capability enables adjuvants to tailor immune outcomes, directing the immune response toward a preferred T-cell subset based on the desired goal of a vaccine or therapy. This strategic modulation of T-cell differentiation through adjuvants plays a vital role in vaccine design, offering a means to optimize protective responses against various pathogens or diseases [38]. Moreover, this effect of adjuvants over T-cell responses also has the potential to impact autoimmune conditions and allergies, highlighting the significance of adjuvant-mediated cytokine regulation in shaping effective immunization strategies and therapeutic interventions.

3.6.3 Cytotoxic T-Cell Activation

Adjuvants also activate cytotoxic T cells (CD8+ T cells), identifying and eliminating infected or aberrant cells. Adjuvants facilitate the intracellular antigen processing performed by APCs and subsequently enhance the presentation of these processed antigens to cytotoxic T cells. This process primes CD8+ T cells for action and ensures their ability to determine cells carrying antigens that signify infection or abnormality accurately.

3.6.4 Cross-Presentation of Exogenous Antigens

Specific adjuvants play a role in facilitating the process of cross-presentation by APCs. This mechanism can enable APCs to present extracellular antigens onto MHC class I molecules, thereby engaging cytotoxic T cells. By bridging the gap between the exogenous nature of the antigens and the MHC class I pathway, cross-presentation facilitated by adjuvants equips the immune system with a robust strategy to recognize and combat intracellular threats effectively.

3.6.5 Inflammation and Immune Cell Recruitment

Adjuvants exert their influence by inducing localized inflammation at the injection site, which catalyzes attracting a diverse array of immune cells to the affected area. This influx of immune cells sets the stage for heightened interactions among APCs, T cells, and other immune cell types. The adjuvant-driven inflammatory microenvironment fosters optimal conditions for efficient antigen uptake, presentation, and activation, resulting in a coordinated and potent immune reaction against intruders or threats.

3.6.6 Memory T-Cell Generation

The memory T cells retain a lasting "memory" of the encountered antigen, enabling a swifter and more robust response upon subsequent encounters with the same pathogen or antigen. Adjuvants enhance the formation of these memory T-cell populations by influencing the activation and differentiation processes during the initial immune response. As a result, the immune system becomes primed to respond rapidly and effectively, significantly shortening the lag time between re-exposure to the antigen and initiating a potent immune reaction.

3.6.7 Enhancement of Antigen Persistence

Specific adjuvants play a role in fostering extended antigen presentation by APCs. These adjuvants facilitate interactions between APCs and T cells, offering a prolonged window for immune cells to engage with the presented antigens. This attribute proves crucial for activating and differentiating specific T-cell subsets that demand sustained exposure to antigens.

3.6.8 Antigen Depot Formation

Adjuvants hold the capacity to generate a localized depot at the site of injection. This phenomenon engenders a gradual and controlled release of both antigens and adjuvants over an extended period. This distinctive mechanism of antigen delivery serves to amplify the activation of T cells, thereby fostering a heightened immune response. Additionally, this sustained antigen and adjuvant exposure plays a pivotal role in cultivating the development of memory T cells.

3.6.9 Induction of Th1 Responses

Adjuvants can stimulate the development of Th1 responses, a distinctive type of immune reaction characterized by the robust production of proinflammatory cytokines such as interferon-gamma (IFN-γ). By inducing the production of IFN-γ and other associated cytokines, Th1 responses increase the ability of the immune system to combat infections caused by microbes that invade and replicate within host cells [2].

Hence, adjuvants help in elevating cellular immune responses through multifaceted processes. By catalyzing the activation of APCs, facilitating T-cell differentiation, promoting the generation of memory T cells, and enhancing the overall effectiveness of T-cell–mediated immunity, adjuvants wield a profound impact on the immune system's capabilities.

3.7 Adjuvants and Innate Immunity

Adjuvants interact extensively with the innate immune system to enhance and shape subsequent adaptive immune responses. In previous segments, we have discussed how adjuvants play a role in enhancing B-cell– and

T-cell-mediated immunity. In this segment, we discuss the role of adjuvants in innate immunity [39].

3.7.1 Phagocytosis and Antigen Processing

Adjuvants can enhance phagocytosis, a vital process wherein immune cells like macrophages and DCs engulf foreign particles, including vaccine antigens. By improving the phagocytic activity of these cells, adjuvants contribute to better antigen capture, processing, and subsequent presentation to adaptive immune cells. When immune cells phagocytose antigens, they break them into smaller fragments and present them on their cell surface using MHC molecules. This antigen presentation is a critical step for activating T cells, which are essential in orchestrating adaptive immune responses. Adjuvants that enhance phagocytosis ensure a more efficient and thorough uptake of vaccine antigens by immune cells, leading to an enhanced pool of antigens available for presentation to T cells.

3.7.2 Complement Activation

Specific adjuvants can activate the complement system, an integral part of the innate immune response. The complement system consists of a group of proteins that, when activated, play various roles in the immune defense against pathogens. Adjuvants that activate the complement system can trigger a cascade of reactions that lead to opsonization, inflammation, and immune cell recruitment. Opsonization involves coating pathogens or particles, including vaccine antigens, with complement proteins. This opsonization "tags" the particles, making them more recognizable to phagocytic cells like macrophages and neutrophils, thus promoting their efficient engulfment and elimination. Additionally, complement activation releases anaphylatoxins, such as C3a and C5a, which attract immune cells to the site of infection or antigen administration, leading to an inflammatory response. Activating the complement system by adjuvants adds another layer of immune stimulation to the vaccine response, enhancing the overall immune activation and the subsequent development of adaptive immune responses.

3.7.3 Induction of Local Inflammation

Adjuvants help in generating a controlled inflammation at the site of vaccine administration. The resulting microenvironment promotes immune cell interaction, activation, and coordination, creating a robust and well-coordinated immune response. By strategically creating this environment, adjuvants ensure optimal recognition, processing, and memory formation against vaccine antigens, leading to enhanced vaccine-induced immunity and long-lasting protection against the targeted pathogen.

3.7.4 Pattern Recognition Receptor Activation

Adjuvants utilize molecules that mimic PAMPs, recognizable structures commonly found on pathogens. These PAMP-like molecules serve as ligands for PRRs on various innate immune cells, including TLRs and NOD-like receptors. When these receptors engage with PAMP-like molecules, they initiate intracellular signaling pathways that activate immune cells. This activation includes the upregulation of pro-inflammatory cytokines, chemokines, and co-stimulatory molecules, all contributing to a heightened immune response. By simulating the recognition of a pathogen, adjuvants harness the innate immune system's ability to recognize and respond to threats, leading to a more potent and sustained adaptive immune response when combined with antigens in vaccines.

3.7.5 Natural Killer (NK) Cell Activation

Adjuvants can indirectly activate natural killer (NK) cells, a subset of innate immune cells critical for identifying and eliminating infected or abnormal cells. While NK cells are primarily associated with direct cytotoxicity, they can also be influenced by the signals generated during adjuvant-driven immune responses. The enhanced inflammatory environment created by adjuvants can release cytokines and chemokines that indirectly activate NK cells. These activated NK cells contribute to the immune response against infections in various ways, including recognizing and eliminating virus-infected and cancer cells.

3.7.6 Activation of Epithelial Cells

Specific adjuvants can activate epithelial cells, which form the protective lining of various body surfaces such as the skin, respiratory tract, and gastrointestinal tract. When activated by adjuvants, these epithelial cells can release a range of factors, including chemokines, cytokines, and other signaling molecules that attract immune cells to the activation site. This immune cell recruitment is a crucial step in amplifying the immune response. The immune cells drawn to the site can include neutrophils, macrophages, and other innate immune cells that contribute to the local immune response. Additionally, the factors released by activated epithelial cells can stimulate immune cells to engage in defense mechanisms, including phagocytosis and cytokine production.

The complex mechanisms by which adjuvants interact with the innate immune system initiate a series of interconnected events culminating in a heightened and effective immune response [40, 41]. Adjuvants act as critical mediators that bridge the gap between the innate and adaptive immune systems, setting the stage for a comprehensive and potent defense against pathogens or antigens. Through their multifaceted engagement with various innate immune components, adjuvants significantly contribute to developing more robust and longer-lasting immunity, thus enhancing the overall efficacy of vaccines and bolstering the body's ability to combat infectious diseases.

3.8 Adjuvants and Mucosal Immunity

Adjuvants can contribute to the reinforcement of mucosal barriers and the preservation of mucosal tissue integrity. These barriers act as the body's first line of defense against invading pathogens, toxins, and other harmful substances. By supporting the maintenance of these barriers, adjuvants play a crucial role in preventing the penetration and colonization of pathogens in mucosal tissues. This not only helps prevent infections but also contributes to the overall health and well-being of the individual by safeguarding the integrity of mucosal surfaces found in the respiratory, gastrointestinal, and urogenital tracts [42].

3.8.1 Enhanced Mucosal Antigen Uptake

Adjuvants significantly influence mucosal immunity by enhancing the uptake of antigens by specialized cells like epithelial cells and M cells in mucosal tissues. These adjuvant-assisted cells are pivotal in capturing antigens from the mucosal environment and transporting them to immune cells, particularly DCs, which initiate robust immune responses. This process leads to the activation of local immune cells, secretion of mucosal-specific antibodies, and the establishment of immune memory. By improving antigen presentation and immune cell activation at mucosal surfaces, adjuvants contribute to a more effective defense against pathogens targeting areas like the respiratory, gastrointestinal, and urogenital tracts.

3.8.2 Secretory IgA Production

Mucosal immune responses are frequently characterized by the production of secretory immunoglobulin A (IgA) antibodies. These antibodies serve as a critical defense mechanism against pathogens at mucosal surfaces, acting as a first line of protection. Adjuvants can amplify the production of secretory IgA antibodies in mucosal tissues to prevent infections at sites like the respiratory, gastrointestinal, and urogenital tracts.

3.8.3 Induction of Tolerance

Specific adjuvants can promote immune tolerance at mucosal sites. Immune tolerance is how the immune system learns to recognize and tolerate harmless antigens, preventing unnecessary and potentially harmful immune responses against them. This mechanism is vital for avoiding exaggerated reactions to environmental substances and managing autoimmune conditions where the immune system mistakenly attacks the body tissues. Adjuvants that support immune tolerance can help maintain a balanced immune response, preventing unwarranted inflammatory reactions and contributing to overall immune system health.

3.8.4 Activation of Dendritic Cells

Adjuvants can stimulate the activation of DCs within mucosal tissues, and these dendritic cells play a central and essential role in initiating immune responses. Once activated by adjuvants, DCs capture antigens from their surroundings and then migrate to nearby lymph nodes, where they present these antigens to T cells. This antigen presentation activates both local and systemic immune responses. Local immune responses at mucosal sites are crucial for preventing infections in those specific areas. These responses include the production of antibodies, particularly secretory IgA, and recruiting other immune cells to the mucosal tissues. In addition, activating DCs by adjuvants leads to the priming of T cells, which are essential for adaptive immune responses. These T cells can circulate throughout the body, contributing to systemic immunity against the targeted pathogen [43].

3.8.5 Recruitment of Effector Cells

Adjuvants can attract immune effector cells, including T cells and B cells, to mucosal tissues. These cells are critical players in immune responses against pathogens. By attracting these immune cells, adjuvants facilitate the establishment of immune reactions within mucosal surfaces, aiding in the clearance of infections and promoting the body's overall defense.

3.8.6 Cross-Presentation at Mucosal Sites

Some adjuvants facilitate the cross-presentation of antigens by DCs within mucosal tissues. Cross-presentation is the process by which DCs present antigens derived from infected or internalized cells to CD8+ cytotoxic T cells. This mechanism is significant for inducing cytotoxic T-cell responses against intracellular pathogens, such as viruses or certain bacteria, that have infected host cells. By promoting cross-presentation, adjuvants enhance the immune system's ability to directly identify and eliminate infected cells.

3.8.7 Improvement of Oral and Nasal Vaccines

Adjuvants hold particular significance for vaccines administered through the oral or nasal routes. These routes provide direct access to mucosal tissues, common entry points for many pathogens. Adjuvants can be critical in enhancing immune responses in these areas, ensuring the body can mount effective defenses against infections targeting mucosal surfaces. By enhancing antigen uptake, promoting immune cell activation, and stimulating the production of mucosal antibodies, adjuvants contribute to a more robust and tailored immune response in the mucosal tissues of the respiratory, gastrointestinal, and urogenital tracts. This is especially relevant for pathogens primarily targeting these areas, such as respiratory viruses or enteric pathogens [44].

Adjuvants help boost mucosal immunity by facilitating essential processes like antigen uptake, local immune response activation, antibody production, and memory cell generation. These functions collectively contribute to developing effective vaccines, particularly against pathogens that target mucosal surfaces as their entry point.

3.9 Adjuvants and Vaccine Efficacy in Specific Populations

Adjuvants are particularly valuable in enhancing vaccine efficacy in populations with less developed, weaker, or compromised immune responses. Here is how adjuvants can improve vaccine efficacy in infant, elderly, and immunocompromised populations [45].

3.9.1 Infant

1) Immature Immune System: Infants have immature immune systems, which may not respond as robustly to vaccines. Adjuvants can enhance immune activation, helping overcome this limitation and generating more substantial and durable immune responses.
2) Reduced Dosing Requirements: Adjuvants allow for smaller vaccine antigen doses, reducing the risk of adverse reactions in infants. This is particularly important given their smaller body size and developing immune system.
3) Improved Immune Memory: Adjuvants aid in the development of memory B cells and memory T cells even in infants, ensuring a more effective response upon subsequent exposures.

3.9.2 Elderly

1) Immunosenescence: Aging leads to immunosenescence, where the immune system becomes less responsive. Adjuvants can overcome this by enhancing immune activation and promoting the generation of memory cells.
2) Enhanced Antibody Responses: Adjuvants stimulate better antibody production, which is crucial for protecting against infections that might be more severe in elderly humans.
3) Reduced Risk of Vaccine Failure: In some cases, standard vaccines may fail to adequately protect the elderly. Adjuvants can improve vaccine effectiveness and decrease the risk of vaccine failure.
4) Protection Against Age-Related Diseases: Adjuvants can aid in developing vaccines for diseases that disproportionately affect the elderly population, such as influenza and pneumococcal infections.

3.9.3 Immunocompromised Individuals

1) Weakened Immune Responses: Individuals with compromised immune systems have reduced capacity to mount effective immune responses. Adjuvants can help overcome this by boosting immune activation and enhancing vaccine efficacy.

2) Tailored Responses: Adjuvants can steer immune responses in specific directions, which is beneficial in cases where a particular immune response is needed to combat a specific pathogen.
3) Reduced Antigen Doses: Adjuvants allow for lower antigen doses, reducing the risk of adverse reactions while promoting immune responses.
4) Protection Against Opportunistic Infections: Immunocompromised individuals are more susceptible to certain infections. Adjuvanted vaccines can better protect against these infections by stimulating robust immune responses.
5) Vaccine Development for Immunocompromised: Adjuvants can be instrumental in the development of vaccines designed explicitly for immunocompromised individuals, offering them the protection that standard vaccines might not provide.

Adjuvants are critical in improving vaccine efficacy in populations with weakened or compromised immune responses, such as infants, aged, and immunocompromised individuals. By enhancing immune activation, promoting memory cell formation, and tailoring immune responses, adjuvants contribute to developing more effective and protective vaccines in these specific populations.

3.10 Conclusion

Adjuvants are critical tools to enhance and fine-tune immune responses, leading to more robust, durable, and targeted immune reactions against pathogens. This understanding holds several critical implications, some of which are summarized below [46].

3.10.1 Enhanced Vaccine Efficacy

Adjuvants amplify and direct immune responses, enabling vaccines to generate more substantial and compelling immunity. By promoting robust antibody production, activating cytotoxic T cells, and generating memory cells, adjuvants significantly improve the ability of vaccines to prevent infections and reduce disease severity.

3.10.2 Tailored Immune Responses

Adjuvants allow for the customization of immune responses. This is crucial for developing vaccines against a wide range of pathogens, as different infections may require different types of immune reactions for optimal protection.

3.10.3 Protection in Vulnerable Populations

Adjuvants are pivotal in improving vaccine efficacy in vulnerable populations, such as infants, the elderly, and immunocompromised individuals. They counteract weakened immune responses and enhance protection in groups traditionally at higher risk of infections.

3.10.4 Reduction in Antigen Doses

Adjuvants enable the use of lower vaccine antigen doses while maintaining or even enhancing immune responses. This reduces the risk of adverse reactions and allows for more efficient vaccine production, distribution, and administration.

3.10.5 Development of Universal Vaccines

Understanding adjuvant-immune system interactions aids in creating universal vaccines that target multiple strains or variants of a pathogen. This potentially revolutionizes vaccine strategies, minimizing the need for frequent updates in response to evolving pathogens.

3.10.6 Management of Emerging Diseases

As new infectious diseases emerge, adjuvants can expedite the development of effective vaccines by rapidly enhancing immune responses against novel pathogens.

3.10.7 Prevention of Epidemics and Pandemics

Rapid vaccine development and deployment, facilitated by adjuvants, are vital tools for controlling outbreaks and preventing the spread of infectious diseases on a global scale.

3.10.8 Public Health Impact

Improving vaccine efficacy through adjuvants reduces disease burden, hospitalizations, and healthcare costs. By preventing infections, vaccines contribute to improved overall public health and well-being.

In conclusion, the intricate interplay between adjuvants and the immune system is a cornerstone of modern vaccine development. It underpins creating safer, more effective, and adaptable vaccines for diverse populations and emerging threats. A thorough understanding of these interactions paves the way for groundbreaking advancements in public health, shaping a world where infectious diseases are better controlled and prevented.

References

1. Lee, A., Wimmers, F., Pulendran, B., Epigenetic adjuvants: durable reprogramming of the innate immune system with adjuvants. *Curr. Opin. Immunol.*, 77, 102189, 2022 Aug, doi: 10.1016/j.coi.2022.102189. Epub 2022 May 16, PMID: 35588691; PMCID: PMC9924100.
2. Awate, S., Babiuk, L.A., Mutwiri, G., Mechanisms of action of adjuvants. *Front. Immunol.*, 4, 114, 2013 May 16, doi: 10.3389/fimmu.2013.00114, PMID: 23720661; PMCID: PMC3655441.
3. Guimarães, L.E., Baker, B., Perricone, C., Shoenfeld, Y., Vaccines, adjuvants and autoimmunity. *Pharmacol. Res.*, 100, 190–209, 2015 Oct, doi: 10.1016/j.phrs.2015.08.003. Epub 2015 Aug 12, PMID: 26275795; PMCID: PMC7129276.
4. Carter, D. and Reed, S.G., Role of adjuvants in modeling the immune response. *Curr. Opin. HIV AIDS.*, 5, 5, 409–13, 2010 Sep, doi: 10.1097/COH.0b013e32833d2cdb, PMID: 20978382; PMCID: PMC3894736.
5. Rasaei, R., Sarodaya, N., Kim, K.S., Ramakrishna, S., Hong, S.H., Importance of Deubiquitination in Macrophage-Mediated Viral Response and Inflammation. *Int. J. Mol. Sci.*, 21, 21, 8090, 2020 Oct 29, doi: 10.3390/ijms21218090, PMID: 33138315; PMCID: PMC7662591.
6. Harryvan, T.J., de Lange, S., Hawinkels, L.J.A.C., Verdegaal, E.M.E., The ABCs of Antigen Presentation by Stromal Non-Professional Antigen-Presenting Cells. *Int. J. Mol. Sci.*, 23, 1, 137, 2021 Dec 23, doi: 10.3390/ijms23010137, PMID: 35008560; PMCID: PMC8745042.
7. Moga, E., Lynton-Pons, E., Domingo, P., The Robustness of Cellular Immunity Determines the Fate of SARS-CoV-2 Infection. *Front. Immunol.*, 13, 904686, 2022 Jun 27, doi: 10.3389/fimmu.2022.904686, PMID: 35833134; PMCID: PMC9271749.

8. Mak, T.W. and Saunders, M.E., Immunity to pathogens, in: *Immune Response*, pp. 641–94, 2006, doi: 10.1016/B978-012088451-3.50024-7, Epub 2007 Sep 2. PMCID: PMC7150295.
9. Chaplin, D.D., Overview of the immune response. *J. Allergy Clin. Immunol.*, 125, 2 Suppl 2, S3–23, 2010 Feb, doi: 10.1016/j.jaci.2009.12.980, PMID: 20176265; PMCID: PMC2923430.
10. Sadarangani, M., Marchant, A., Kollmann, T.R., Immunological mechanisms of vaccine-induced protection against COVID-19 in humans. *Nat. Rev. Immunol.*, 21, 8, 475–484, 2021 Aug, doi: 10.1038/s41577-021-00578-z. Epub 2021 Jul 1, PMID: 34211186; PMCID: PMC8246128.
11. Gaimann, M.U., Nguyen, M., Desponds, J., Mayer, A., Early life imprints the hierarchy of T-cell clone sizes. *Elife*, 9, e61639, 2020 Dec 21, doi: 10.7554/eLife.61639, PMID: 33345776; PMCID: PMC7870140.
12. Retamal-Díaz, A., Covián, C., Pacheco, G.A., Castiglione-Matamala, A.T., Bueno, S.M., González, P.A., Kalergis, A.M., Contribution of Resident Memory CD8+ T-cells to Protective Immunity Against Respiratory Syncytial Virus and Their Impact on Vaccine Design. *Pathogens*, 8, 3, 147, 2019 Sep 11, doi: 10.3390/pathogens8030147, PMID: 31514485; PMCID: PMC6789444.
13. Ishii-Mizuno, Y., Umeki, Y., Onuki, Y., Watanabe, H., Takahashi, Y., Takakura, Y., Nishikawa, M., Improved sustained release of antigen from immunostimulatory DNA hydrogel by electrostatic interaction with chitosan. *Int. J. Pharm.*, 516, 1–2, 392–400, 2017 Jan 10, doi: 10.1016/j.ijpharm.2016.11.048, Epub 2016 Nov 21. PMID: 27884715.
14. van Aalst, S., Ludwig, I.S., van Kooten, P.J.S., van der Zee, R., van Eden, W., Broere, F., Dynamics of APC recruitment at the site of injection following injection of vaccine adjuvants. *Vaccine*, 35, 12, 1622–1629, 2017 Mar 14, doi: 10.1016/j.vaccine.2017.02.005, Epub 2017 Feb 17. PMID: 28222998.
15. Hutchison, S., Benson, R.A., Gibson, V.B., Pollock, A.H., Garside, P., Brewer, J.M., Antigen depot is not required for alum adjuvanticity. *FASEB J.*, 26, 3, 1272–9, 2012Mar, doi: 10.1096/fj.11-184556, Epub 2011 Nov 21. PMID: 22106367; PMCID: PMC3289510.
16. Mohan, T., Zhu, W., Wang, Y., Wang, B.Z., Applications of chemokines as adjuvants for vaccine immunotherapy. *Immunobiology*, 223, 6–7, 477–485, 2018 Jun-Jul, doi: 10.1016/j.imbio.2017.12.001, Epub 2017 Dec 8. PMID: 29246401; PMCID: PMC5894355.
17. García-Piñeres, A., Hildesheim, A., Dodd, L., Kemp, T.J., Williams, M., Harro, C., Lowy, D.R., Schiller, J.T., Pinto, L.A., Cytokine and chemokine profiles following vaccination with human papillomavirus type 16 L1 Virus-like particles. *Clin. Vaccine Immunol.*, 14, 8, 984–9, 2007 Aug, doi: 10.1128/CVI.00090-07. Epub 2007 Jun 27. PMID: 17596432; PMCID: PMC2044489.
18. Wilkins, A.L., Kazmin, D., Napolitani, G., Clutterbuck, E.A., Pulendran, B., Siegrist, C.A., Pollard, A.J., AS03- and MF59-Adjuvanted Influenza Vaccines in Children. *Front. Immunol.*, 8, 1760, 2017 Dec 13, doi: 10.3389/fimmu.2017.01760, PMID: 29326687; PMCID: PMC5733358.

19. Calabro, S., Tortoli, M., Baudner, B.C., Pacitto, A., Cortese, M., O'Hagan, D.T., De Gregorio, E., Seubert, A., Wack, A., Vaccine adjuvants alum and MF59 induce rapid recruitment of neutrophils and monocytes that participate in antigen transport to draining lymph nodes. *Vaccine*, 29, 9, 1812–23, 2011 Feb 17, doi: 10.1016/j.vaccine.2010.12.090, Epub 2011 Jan 6. PMID: 21215831.
20. Swain, S.L., McKinstry, K.K., Strutt, T.M., Expanding roles for $CD4^+$ T-cells in immunity to viruses. *Nat. Rev. Immunol.*, 12, 2, 136–48, 2012 Jan 20, doi: 10.1038/nri3152, PMID: 22266691; PMCID: PMC3764486.
21. Cioncada, R., Maddaluno, M., Vo, H.T.M., Woodruff, M., Tavarini, S., Sammicheli, C., Tortoli, M., Pezzicoli, A., De Gregorio, E., Carroll, M.C., D'Oro, U., Piccioli, D., Vaccine adjuvant MF59 promotes the intranodal differentiation of antigen-loaded and activated monocyte-derived dendritic cells. *PLoS One*, 12, 10, e0185843, 2017 Oct 31, doi: 10.1371/journal.pone.0185843, PMID: 29088224; PMCID: PMC5663329.
22. Ivanov, K., Garanina, E., Rizvanov, A., Khaiboullina, S., Inflammasomes as Targets for Adjuvants. *Pathogens*, 9, 4, 252, 2020 Mar 30, doi: 10.3390/pathogens9040252, PMID: 32235526; PMCID: PMC7238254.
23. Apostólico Jde, S., Lunardelli, V.A., Coirada, F.C., Boscardin, S.B., Rosa, D.S., Adjuvants: Classification, Modus Operandi, and Licensing. *J. Immunol. Res.*, 2016, 1459394, 2016, doi: 10.1155/2016/1459394, Epub 2016 May 4. PMID: 27274998; PMCID: PMC4870346.
24. Nooraei, S., Sarkar Lotfabadi, A., Akbarzadehmoallemkolaei, M., Rezaei, N., Immunogenicity of Different Types of Adjuvants and Nano-Adjuvants in Veterinary Vaccines: A Comprehensive Review. *Vaccines (Basel)*, 11, 2, 453, 2023 Feb 16, doi: 10.3390/vaccines11020453, PMID: 36851331; PMCID: PMC9962389.
25. Lindblad, E.B. and Duroux, L., Mineral adjuvants, in: *Immunopotentiators in Modern Vaccines*, pp. 347–75, Amsterdam, Netherlands, 2017, Epub 2017 Jan 13, PMCID: PMC7149584, doi: 10.1016/B978-0-12-804019-5.00018-9.
26. Madera, R., Burakova, Y., Shi, J., Emulsion Adjuvants for Use in Veterinary Vaccines. *Methods Mol. Biol.*, 2412, 247–253, 2022, doi: 10.1007/978-1-0716-1892-9_11, PMID: 34918248.
27. Haensler, J., Manufacture of Oil-in-Water Emulsion Adjuvants. *Methods Mol. Biol.*, 1494, 165–180, 2017, doi: 10.1007/978-1-4939-6445-1_12, PMID: 27718193.
28. Moni, S.S., Abdelwahab, S.I., Jabeen, A., Elmobark, M.E., Aqaili, D., Ghoal, G., Oraibi, B., Farasani, A.M., Jerah, A.A., Alnajai, M.M.A., Mohammad Alowayni, A.M.H., Advancements in Vaccine Adjuvants: The Journey from Alum to Nano Formulations. *Vaccines*, 11, 11, 1704, 2023. https://doi.org/10.3390/vaccines11111704 PMID: 38006036
29. Madera, R., Burakova, Y., Shi, J., Emulsion Adjuvants for Use in Veterinary Vaccines. *Methods Mol. Biol.*, 2412, 247–253, 2022, doi: 10.1007/978-1-0716-1892-9_11, PMID: 34918248.

30. Haensler, J., Manufacture of Oil-in-Water Emulsion Adjuvants. *Methods Mol. Biol.*, 1494, 165–180, 2017, doi: 10.1007/978-1-4939-6445-1_12., PMID: 27718193.
31. Prior, J.T., Davitt, C., Kurtz, J., Gellings, P., McLachlan, J.B., Morici, L.A., Bacterial-Derived Outer Membrane Vesicles are Potent Adjuvants that Drive Humoral and Cellular Immune Responses. *Pharmaceutics*, 13, 2, 131, 2021 Jan 20, doi: 10.3390/pharmaceutics13020131, PMID: 33498352; PMCID: PMC7909432.
32. McElrath, M.J., Adjuvants: tailoring humoral immune responses. *Curr. Opin. HIV AIDS*, 12, 3, 278–284, 2017 May, doi: 10.1097/COH.0000000000000365, PMID: 28257301; PMCID: PMC5510533.
33. Xu, W. and Banchereau, J., The antigen presenting cells instruct plasma cell differentiation. *Front. Immunol.*, 4, 504, 2014 Jan 6, doi: 10.3389/fimmu.2013.00504, PMID: 24432021; PMCID: PMC3880943.
34. Ko, E.J., Lee, Y.T., Kim, K.H., Jung, Y.J., Lee, Y., Denning, T.L., Kang, S.M., Effects of MF59 Adjuvant on Induction of Isotype-Switched IgG Antibodies and Protection after Immunization with T-Dependent Influenza Virus Vaccine in the Absence of CD4+ T-cells. *J. Virol.*, 90, 15, 6976–6988, 2016 Jul 11, doi: 10.1128/JVI.00339-16, PMID: 27226368; PMCID: PMC4944285.
35. Kao, D., Lux, A., Schaffert, A., Lang, R., Altmann, F., Nimmerjahn, F., IgG subclass and vaccination stimulus determine changes in antigen specific antibody glycosylation in mice. *Eur. J. Immunol.*, 47, 12, 2070–2079, 2017 Dec, doi: 10.1002/eji.201747208, Epub 2017 Aug 29. PMID: 28771702.
36. Chijns, V., Fernández-Tejada, A., Barjaktarović, Z., Bouzalas, I., Brimnes, J., Chernysh, S., Gizurarson, S., Gursel, I., Jakopin, Z., Lawrenz, M., Nativi, C., Paul, S., Pedersen, G.K., Rosano, C., Ruiz-de-Angulo, A., Slütter, B., Thakur, A., Christensen, D., Lavelle, E.C., Modulation of immune responses using adjuvants to facilitate therapeutic vaccination. *Immunol. Rev.*, 296, 1, 169–190, 2020 Jul, doi: 10.1111/imr.12889, Epub 2020 Jun 28. PMID: 32594569; PMCID: PMC7497245.
37. Rapaka, R.R., Cross, A.S., McArthur, M.A., Using Adjuvants to Drive T-cell Responses for Next-Generation Infectious Disease Vaccines. *Vaccines (Basel)*, 9, 8, 820, 2021 Jul 24, doi: 10.3390/vaccines9080820, PMID: 34451945; PMCID: PMC8402546.
38. McAleer, J.P. and Vella, A.T., Educating CD4 T-cells with vaccine adjuvants: lessons from lipopolysaccharide. *Trends Immunol.*, 31, 11, 429–35, 2010 Nov, doi: 10.1016/j.it.2010.08.005. Epub 2010 Sep 27, PMID: 20880743; PMCID: PMC2967613.
39. Coffman, R.L., Sher, A., Seder, R.A., Vaccine adjuvants: putting innate immunity to work. *Immunity*, 33, 4, 492–503, 2010 Oct 29, doi: 10.1016/j.immuni.2010.10.002. PMID: 21029960; PMCID: PMC3420356.
40. Akira, S., Innate immunity and adjuvants. *Philos. Trans. R Soc. Lond. B Biol. Sci.*, 366, 1579, 2748–55, 2011 Oct 12, doi: 10.1098/rstb.2011.0106. PMID: 21893536; PMCID: PMC3146784.

41. Kuroda, E., Coban, C., Ishii, K.J., Particulate adjuvant and innate immunity: past achievements, present findings, and future prospects. *Int. Rev. Immunol.*, 32, 2, 209–20, 2013 Apr, doi: 10.3109/08830185.2013.773326, PMID: 23570316; PMCID: PMC3632271.
42. Aoshi, T., Modes of Action for Mucosal Vaccine Adjuvants. *Viral Immunol.*, 30, 6, 463–470, 2017 Jul/Aug, doi: 10.1089/vim.2017.0026. Epub 2017 Apr 24. PMID: 28436755; PMCID: PMC5512297.
43. Correa, V.A., Portilho, A.I., De Gaspari, E., Vaccines, adjuvants and key factors for mucosal immune response. *Immunology*, 167, 2, 124–138, 2022 Oct, doi: 10.1111/imm.13526, Epub 2022 Jul 12. PMID: 35751397.
44. Lawson, L.B., Norton, E.B., Clements, J.D., Defending the mucosa: adjuvant and carrier formulations for mucosal immunity. *Curr. Opin. Immunol.*, 23, 3, 414–20, 2011 Jun, doi: 10.1016/j.coi.2011.03.009, Epub 2011 Apr 19. PMID: 21511452.
45. Nanishi, E., Dowling, D.J., Levy, O., Toward precision adjuvants: optimizing science and safety. *Curr. Opin. Pediatr.*, 32, 1, 125–138, 2020 Feb, doi: 10.1097/MOP.0000000000000868, PMID: 31904601; PMCID: PMC6970548.
46. Facciolà, A., Visalli, G., Laganà, A., Di Pietro, A., An Overview of Vaccine Adjuvants: Current Evidence and Future Perspectives. *Vaccines (Basel)*, 10, 5, 819, 2022 May 22, doi: 10.3390/vaccines10050819, PMID: 35632575; PMCID: PMC9147349.

4

Antigen Selection and Design

Pankti C. Balar[1], Anjali P. Bedse[2], Vivek P. Chavda[3*], Chukwuebuka E. Umeyor[4,5], Prafull Kolekar[4], Brian O. Ogbonna[6,7], Ankita Anure[4], Daniel U. Eze[8], Payal Dodiya[1] and Vandana B. Patravale[4†]

[1]Pharmacy Section, L.M. College of Pharmacy, Ahmedabad, India
[2]Department of Pharmaceutics, K K Wagh College of Pharmacy, Nashik, Affiliated to Dr. Babasaheb Ambedkar Technological University, Lonere, Raigad, Maharashtra, India
[3]Department of Pharmaceutics and Pharmaceutical Technology, L.M. College of Pharmacy, Ahmedabad, India
[4]Department of Pharmaceutical Sciences and Technology, Institute of Chemical Technology, Matunga, Mumbai, Maharashtra, India
[5]Department of Pharmaceutics and Pharmaceutical Technology, Faculty of Pharmaceutical Sciences, Nnamdi Azikiwe University, Awka, Anambra State, Nigeria
[6]Department of Clinical Pharmacy and Pharmacy Management, Faculty of Pharmaceutical Sciences, Nnamdi Azikiwe University, Awka, Nigeria
[7]Department of Clinical Pharmacy and Pharmacy Practice, Faculty of Pharmacy, David Umahi Federal University of Health Sciences, Uburu, Nigeria
[8]Department of Pharmaceutical Microbiology and Biotechnology, Faculty of Pharmacy, David Umahi Federal University of Health Sciences, Uburu, Nigeria

Abstract

Different vaccines have been designed and developed for the prevention of infectious diseases. Whether for immunization against viral or bacterial infections, antigens are very critical in the formulation development of effective vaccines. The antigenic material incorporated in vaccines could be attenuated or live molecules, and they are expected to induce strong immune response against assault by an invading or extraneous microbe. Sometimes, the immunostimulatory activity of an antigen could be boosted by the addition of an adjuvant, but both constituents

**Corresponding author*: Vivek.chavda@lmcp.ac.in
†*Corresponding author*: vb.patravale@ictmumbai.edu.in

Vivek P. Chavda and Vasso Apostolopoulos (eds.) Emerging Pathways of Vaccine Adjuvants: A Nonspecific Stimulant of the Immune System, (73–106) © 2025 Scrivener Publishing LLC

of a vaccine must be compatible and synergistic in order to produce the desired immune reactions. In this chapter, we will discuss the various types of vaccines and protocols for the design and selection of antigens for the preparation of vaccines.

Keywords: Antigen, vaccine, pathogen, DNA, viral, recombinant, immunogenicity

4.1 Introduction

Immunizations are being administered routinely throughout life to prevent a variety of diseases that are targeted by vaccinations. Immunization endeavors successfully accomplished the worldwide eradication of smallpox, as well as the elimination of polio, measles, and rubella. The children's vaccine series, which includes DTP (diphtheria-tetanus-pertussis), polio, MMR (measles, mumps, and rubella), Hib, hepatitis B, and varicella vaccines, is intended to prevent diseases, prevent premature mortality, and reduce direct medical expenses. Vaccines have not only reduced infections and hospitalizations among the targeted populations but also provided additional advantages by reducing transmission from vaccinated individuals to other groups [1]. The ability to create herd immunity is the added benefit of vaccination from a community health perspective. Immunizations protect individuals without a vaccination, who may be too young, too vulnerable, or too immunosuppressed to receive vaccinations, by halting the spread of the infectious agent if a significant enough percentage of the population is immunized [2]. The majority of existing vaccine methods exhibit low immunogenicity, provide limited protection, or carry the risk of regaining pathogenic properties. There is a requirement for the development of vaccinations that can be preserved for a long period of time, are easy to administer, and ensure a strong and prolonged immune response [3]. Immunization offers lifetime protection against certain diseases in addition to halting the spread of infection during childhood. However, due to immunological barriers like insufficient immune response and weak immunological memory against vaccines, the scientific community still faces difficulties in creating the best vaccines against a variety of infectious diseases, including the plague, TB, malaria, HIV, and SARS-CoV-2. In addition to these challenges, vaccine safety issues like waning immunity caused by tetanus and diphtheria toxoids along with acellular pertussis, adverse effects in a population with rare genetic disorders, and systemic and local reactogenicity caused by these diseases together with whole cell pertussis have been unacceptable [4, 5]. This increases the impact of the challenge to solve the vaccine problem for emerging or reemerging disease threats. These challenges demand new methods that can improve the understanding of how the immune system responds to vaccinations and offer novel approaches that boost immunity without affecting effectiveness or safety [4]. An adjuvanted vaccine will tend to have a higher, earlier, and longer-lasting immune response than a non-adjuvanted vaccine and save countless lives every year from fatal infectious diseases [6]. Adjuvants are essential elements that increase the efficacy of vaccination practices through many different actions, especially acting as carriers, depots, and stimulators of immune responses. The addition of adjuvants is particularly useful for vaccines used in the elderly due to the physiological phenomenon of immunosenescence occurring in this category of subjects, which is responsible for the reduction of immune responses after natural infections or artificial stimuli (vaccinations) [7]. Adjuvants as immunostimulants activate APCs by targeting specific PRRs, leading to enhanced antigen-presenting and co-stimulatory signaling, which results in adaptive immune enhancement. Adjuvants as delivery systems target antigens to APCs or lymph nodes by prolonging the bioavailability of antigens to enhance the uptake and presentation of antigens by APCs, resulting in an enhanced adaptive immune response [8].

This chapter would cover the importance of vaccines, challenges in vaccine development, role of adjuvants, rationale for novel formulation strategies, emerging technologies, optimization of adjuvant–antigen interactions, delivery systems, immunological considerations, safety, and regulatory considerations, clinical translation and commercialization, case studies, and future directions. The chapter would provide valuable insights into the latest advancements and challenges in vaccine development.

4.2 Types of Antigens Used in Vaccines

4.2.1 Whole Inactivated Pathogens

Vaccines made of entire microorganisms that have been chemically or physically inactivated for example, by heat, radiation, or chemicals are referred to as whole inactivated pathogens. These vaccinations contain all of the pathogen's surface antigens, which elicit an immunological response from the body. By inactivating the pathogen, its antigenic qualities are preserved, and it is rendered incapable of causing illness. After being administered, the immune system views these antigens as alien and launches an attack, creating pathogen-specific antibodies and memory cells.

This reaction primes the immune system to identify and combat the pathogen's future infections [9]. The inactivated polio vaccine and the whole-cell pertussis vaccine are two vaccinations that use whole-inactivated microorganisms. Although these vaccinations are successful at eliciting immunological responses, several doses may be necessary to sustain long-term protection [10].

4.2.2 Live Attenuated Vaccine

Live attenuated vaccines include viruses that are less virulent than their wild-type counterparts, yet they are nonetheless able to replicate within the host. Genetic modification or repeated passage in cell culture can be used to accomplish this attenuation. Vaccines that have been live-attenuated resemble real diseases in that they cause strong, persistent immune responses. They promote humoral and cellular immunity, offering strong defense against the infection in the future [11]. The oral polio vaccine and the MMR vaccine are two examples of live attenuated vaccinations. Despite their great efficacy, these vaccinations can be dangerous for people with reduced immune systems, and they should not be administered to pregnant women or people with impaired immune systems [12].

4.2.3 Viral Subunit Vaccines

Vaccines containing pure viral components, usually proteins or glycoproteins that can trigger an immune response, are known as viral subunit vaccines. Since these vaccinations do not include live virus particles, there is no chance of illness. Subunit vaccines target certain viral antigens to reduce the possibility of side effects while yet eliciting protective response. For those whose underlying medical issues prevent them from receiving live vaccinations, they are frequently safer substitutes [13]. The hepatitis B vaccination, which includes the hepatitis B surface antigen, and the human papillomavirus (HPV) vaccine, which contains virus-like particles (VLPs) made of HPV proteins, are two examples of viral subunit vaccines. Subunit vaccines provide a secure and efficient means of preventing viral infections, even if they could need adjuvants or several doses to maximize immunogenicity. Furthermore, developments in recombinant DNA technology have made it easier to produce subunit vaccines with better safety and effectiveness characteristics [14]. Figure 4.1 shows various antigenic constituents of vaccines.

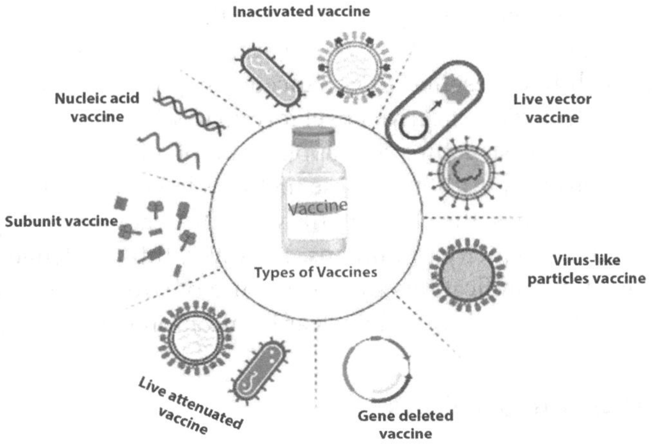

Figure 4.1 Pictorial representation of antigens used in the development of various vaccines. Figure was adapted from [11] under the terms of the Creative Commons Attribution License (CC BY).

4.2.4 Conjugate Vaccines

In developed and developing regions of the world, conjugate vaccines have been used to make remarkable impact in the control of infectious diseases especially those caused by bacteria. Conjugate vaccines are formulated by chemical conjugation of a protein molecule isolated from bacteria with a miniscule amount of polysaccharide constituent of the bacterial cell membrane [14]. Usually, the polysaccharide material enables the bacteria to evade recognition and processing by antigen-presenting cells (APCs) and clearance by the immune cells. Historically, *Haemophilus influenzae* type b (Hib) conjugate vaccine was the first human conjugate vaccine approved and licensed for use in the USA in 1987. Others include pneumococcal and meningococcal conjugate vaccines. Following the inclusion of conjugate vaccines in paediatric immunization schedules, the administration of prescription antibacterial agents for infections caused by encapsulated bacteria have declined significantly due to the success recorded in the prevention of these infections [15].

4.2.5 DNA Vaccines

The clinical importance of DNA vaccines came into global reckoning during the latter part of the 20th century when Tang and Johnston attempted the transdermal delivery of human growth hormone using murine models as gene therapy [16]. Since this pioneering work, there has been increasing interest in DNA vaccines owing to technological advances, which have resulted to enhanced gene optimization, improved DNA delivery strategies, effective novel formulation techniques, enhanced design of RNA microstructure, and application of adjuvants in DNA vaccine formulations [17]. Although these methods led to the development of effective DNA vaccines, the clinical utility of DNA vaccines is impeded by decreased immunogenicity in humans. However, this challenge could be solved by genetic modifications such as the use of strong promoters, Toll-like receptor (TLR) agonists, and nuclear localization peptide genes [18].

4.2.6 Other Antigen Types

Toxoid vaccines are prepared using attenuated toxins, which induce humoral immunity during vaccination or immunization. The selected toxins are responsible for the symptoms of the disease. For instance, tetanus toxoid vaccine could be prepared by purifying tetanus toxin and inactivating it with formaldehyde to formulate a toxoid. The vaccines do not target bacterial cells directly but target infectious bacterial molecules, which compromise immune responses. They require booster immunization or revaccination to enhance individual immunity. Another example of toxoid vaccine is diphtheria vaccine [19].

Blood group antigens have been characterized as proteins (single-pass, multi-pass, or glycosylphosphatidylinositol) or sugar (glycan or carbohydrate) moieties attached to specific components of the red blood cell membrane. They act as surface biomarkers for the ABO (blood grouping system, as sugars) and Rh blood group (as proteins). Individual DNA of the red blood cells determines the type of enzyme and sugar or protein antigens present in the cells [20].

4.2.7 Considerations for Antigen Selection

4.2.7.1 Specificity and Immunogenicity

Antigens are biological compounds that illicit immune responses and may be self-antigens or auto-antigens. Immunogenicity encompasses the induction of humoral and/or cell-mediated immune response. However, the targeted combination with the final products of immune response like the antibodies and/or T-cell surface receptors underscores the concept of antigenicity. Antigen selects lymphocytes for activation from a population or community of cells designed and pre-structured to generate specific antibodies. The antibody-forming cells are cell-specific. They interface and express receptors on the cell-surface, which are capable of interacting with and binding to foreign antigens [21]. In clonal selection, the selection of antigens for proliferation is dependent on the cells with the ability to reorganize only those cells that recognize the antigens. A well-coordinated, well distributed, and equally distributed stimulation elicited by the whole antigen is very necessary for the development of immunity [22].

The specific natures of antigens to designated immune target points or memory are fundamental to the concept of adaptive immunity that allows mammals to actively and promptly respond to infections that may be recurrent. This forms the basis for the treatment of recurrent infections through the induction of antigen-specific memory. However, its response to self is the basis for the advent of allergic and autoimmune diseases. Improvement in these principles gave rise to target-specific interventions like antigen-specific biologic and immunotherapy [23]. Considerations for primary antibodies that bind to the antigens of interest directly include host species, antigen of interest, forgiveness of the molecules, types of molecules, class, clonality, composition, and antibody labeling. The reactions itself is dependent on the bonding formation between the antigen and antibody non-covalent bonding which may be multiple. Examples include Van der Waals forces, electrostatic bonds, and hydrophobic and hydrogen bonding [24].

4.2.7.2 Target Pathogen Life Cycle Stage

Pathogens are microorganisms that can cause diseases. The kinds of pathogenic organisms include viruses, bacteria, protozoa, and fungi. All pathogens transit a lifecycle, which takes the organism from tender to adult viable to reproduce more adults. At the onset, the first step includes intracellular pathogen binding to the surface of the host target cell. In a virus, binding is mediated through the association of a viral surface protein with the specific receptor on the surface of the host cell. The stages of infection include incubation, and prodromal, which are the early stages of illness. This is followed by the middle stage which comprises the stage of decline and convalescence. The later stage comprises of incubation stage, which is the period from exposure to a pathogen until the manifestation of symptoms commences. The organisms then spread to the disease site and targeted organs and consequently shed into the surrounding environment. It overcomes the immunity of the host causing damage in the form of disease. Transmission of the organism from birds and other animals to man may be associated with mutation, homologous recombination, and co-divergence [25].

A good understanding of these life cycle dynamics underscores the principles and techniques for the diagnostic tools and those used for the treatment and prevention of the associated diseases. An evolving example is the use of bacteriophages in phage therapy for the treatment of infections associated with pathogenic bacteria in this era of microbial resistance. Formulations with silver nanoparticles with minimal toxicity, good tenability, and high penetration profile with host cells are essential considerations in new drug development especially in antiviral therapy [26].

4.2.7.3 Safety and Stability

When antigen selection is based on the degree or extent of expression during infection, it facilitates vaccine development for an intracellular pathogen. The use of differentially stable derivatives shows that antigen stability controls antigen proteolysis, antigen processing in APCs, T-cell stimulation, and kinetics of expression of T-cell determinants. Alterations of the stabilizing parameters of the protein interaction network, natural mutations modulate T cells subsequently and help the microbes to escape the immune response. In vaccines, it is expedient that antigens be absorbed into the adjuvant [27]. The resultant vaccine and adjuvant structure of the modified adsorption is essential in understanding the immunogenicity of the vaccine and adjuvant. Structural changes, pH, denaturant, redox medium, and variations in temperatures of the proteins are major determinants of the degradation process or its ability to maintain its conformational stability. Adsorptions of the proteins into aluminum compounds facilitate the stability of the antigen, whereas the non-adsorbed ones can promote the vaccine degradation process. In the 21st century, recombinant proteins are predominantly being used in place of whole organisms for prevention and treatment of immune-related cases [28].

4.3 Antigen Design Strategies

4.3.1 Recombinant Protein Engineering (RPE)

Recombinant protein engineering (RPE) is an effective method for designing antigens for use in vaccine development and diagnostics, among other applications. Important RPE techniques take into account the following factors:

4.3.1.1 *Selection of Antigenic Regions*

Recognizing the antigenic regions associated with the target pathogen is often crucial in developing an antigen. Bioinformatics analysis, experimental techniques like epitope mapping, or a mix of the two can be used to accomplish this. Antigenic areas can be chosen to produce recombinants when they have been identified [29].

4.3.1.2 *Gene Cloning*

The gene sequences that code for the antigenic regions are either intentionally produced or amplified using PCR (polymerase chain reaction). Following that, these gene sequences are cloned into an appropriate expression vector [29].

4.3.1.3 *Vector Selection*

The vector is essential and is dependent on several parameters, including the target protein folding, post-translational modifications, downstream uses, and the host organism (e.g., bacteria, yeast, and mammalian cells) for expression [30].

4.3.1.4 *Codon Optimization*

Codon optimization can be used to increase protein expression levels in the specified expression system. This entails substituting uncommon codons with more commonly used ones in the host organism in order to increase translation efficiency [30].

4.3.1.5 *Fusion Tags*

Fusion tags, such as polyhistidine (His-tag), glutathione-S-transferase (GST), or maltose-binding protein (MBP), can be added to recombinant proteins to help in purification and detection [31].

4.3.1.6 *Protein Refolding*

If the recombinant protein is produced as inclusion bodies (insoluble aggregates), then protein refolding methods can be used to produce physiologically active proteins. This entails denaturing the protein, followed by progressive renaturation in controlled circumstances [32].

4.3.1.7 *Post-Translational Modifications*

In certain circumstances, post-translational changes like as glycosylation, phosphorylation, or lipidation are required for antigenicity or immunogenicity. Recombinant protein engineering may require altering the expression system to accomplish such alterations [32].

4.3.1.8 *Structural Stabilization*

Engineering tactics such as disulfide bond formation and the introduction of stabilizing mutations can be used to improve the recombinant protein's stability and structural integrity [33].

4.3.1.9 *Multimerization*

In some circumstances, antigens may need to be given in a multimeric form to increase immunogenicity. Recombinant protein engineering can be utilized to make multimeric antigen constructs [34–36].

Pictorial representation of methods for the selection of antigens is shown in Figure 4.2. Recombinant protein engineering provides a versatile

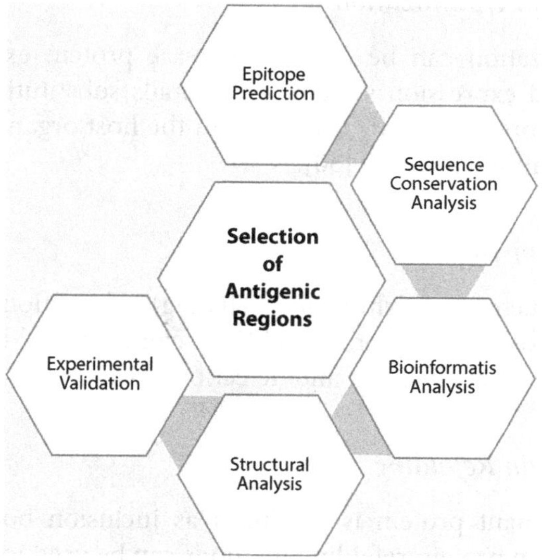

Figure 4.2 Flow diagram showing protocols for the selection of antigens.

and adaptable method for creating antigens tailored to specific purposes such as vaccinations, diagnostics, and treatments.

4.3.2 Peptide Optimization

Peptide optimization is an important feature of antigen design, notably in vaccine development and immunotherapy.

4.3.2.1 Identification of Immunogenic Epitopes

The first step is finding the areas of the target protein with immunogenic epitopes. This can be accomplished using various computational tools, experimental methods like peptide mapping, or a literature study [37].

4.3.2.2 Selection of Conserved Regions

Peptide manufacturers choose to target conserved parts of the protein sequence because they are more likely to elicit widespread immune responses across many strains or species.

4.3.2.3 Length Optimization

When it comes to the presentation of antigens and the immune response, peptide length is vital. Peptides of 8 to 20 amino acids are often used because they have the ability to attach to major histocompatibility complex (MHC) molecules and activate T lymphocytes [38].

4.3.2.4 Modification of Amino Acid Residues

To improve the immunogenicity, binding affinity to MHC molecules, or peptide stability, amino acid insertions, deletions, or substitutions might be made. A non-native amino acid that is resistant to proteolysis can be used to replace sensitive residues, for instance, to lengthen the half-life of peptides [39].

4.3.2.5 Conjugation to Carrier Proteins

Peptides can be combined with carrier proteins (such as bovine serum albumin or keyhole limpet hemocyanin) to improve immunogenicity by stimulating antibody responses and helping T cells [40].

4.3.3 Reverse Vaccinology

Reverse vaccinology is an antigen design process that uses genomic analysis to uncover prospective vaccine candidates rather than standard methods of extracting and characterizing antigens directly from the disease. Advances in genetics and bioinformatics made this method practicable (Figure 4.3) [41].

4.3.3.1 *Genome Sequencing*

The first step is to sequence the genome of the pathogen of interest. This provides a large quantity of genetic data, which may be evaluated to discover prospective vaccine candidates.

4.3.3.2 *General Process Followed for Reverse Vaccinology Platform-Based Modification*

Memory B cells or plasmablasts from seropositive patients with high serum functional titers are commonly employed to generate human monoclonal

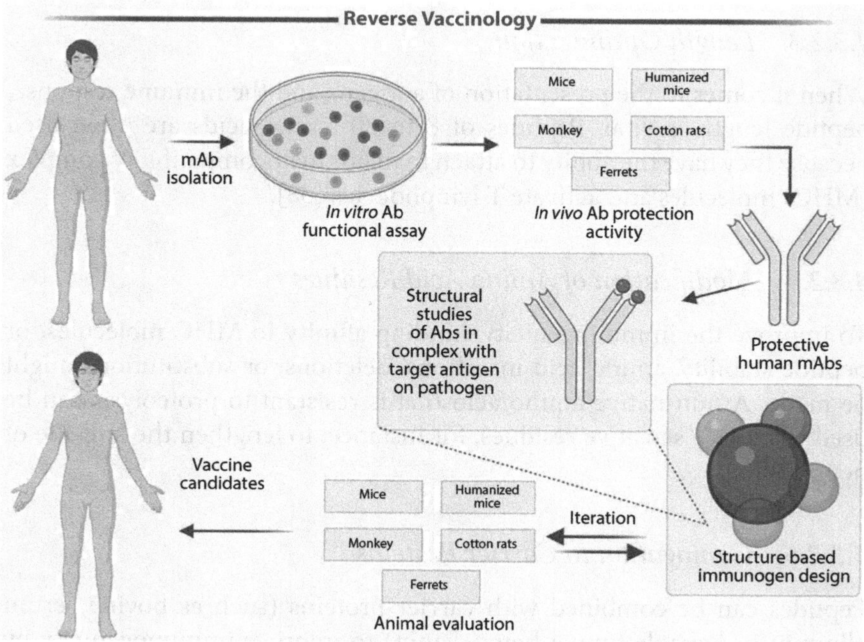

Figure 4.3 Description of the mechanism of reverse vaccinology in the design of antigens.

antibodies (mAbs), which are obtained via direct functional screening or antigen selection methods. The mAbs are tested for *in vivo* protective activity in relevant animal models (for example, mice and ferrets for influenza virus, cotton rats for RSV, and monkeys for HIV). Protective mAbs are examined in conjunction with their pathogen targets (e.g., Env molecules for numerous viruses and structural data used to assist immunogen creation). The immunogens are subsequently tested in animal models. Iterative improvements are often expected in immunogen design before moving on to vaccine candidates for human testing [42].

4.4 Adjuvants: Mechanism of Action and Types

4.4.1 The Rationale for Using Adjuvants

The first known use of adjuvants was reported in 1920 by the French scientist Gaston Ramon. He found that adding aluminum salts to vaccines increases their effectiveness. The Latin word adjuvare, which means "to help," is where the word "adjuvant" first appeared. In general, adjuvants are not immunogenic, but they can change the immune responses when combined with the vaccines. It may result in a lowering of the dosage needed for vaccination, and it may also induce the duration of immunological memory. To assess the potential of preventing the spread of disease, vaccinations are usually prepared with appropriate adjuvants that increase the immune responses to the given vaccine antigen. Adjuvants also play a critical role in directing humoral immunity and cell-mediated immune responses to produce pathogen-specific immune responses [43]. Adjuvants are currently used to increase the immunological response to the administered vaccine, boost the antibody response, and increase the number of recipients who received the vaccination; to raise seroconversion rates in people whose responsiveness has been compromised by disease, aging, or therapeutic interventions [44]; and to decrease both the dose and the number of booster shots of vaccine antigens as the adjuvant's capacity to provide similar responses with considerably smaller doses of vaccine antigen may be essential in areas where there are very few resources to manufacture vaccines and mass immunization is required [45]. The goal of using an adjuvant in vaccine formulation is to achieve qualitative modification of the immune response. During the development of vaccine, adjuvants are more frequently utilized to modify immune response types that vaccine antigens alone are unable to effectively induce. In preclinical and clinical studies, adjuvants are used in the following ways: (i) to produce functionally appropriate immune responses (e.g., humoral or cellular, Th1, Th2, and Th17); additionally, it has been observed that balanced Th1/Th2/Th17 responses lengthen T-cell responses and prolong mouse survival [46]; (ii) for enhancing long-term memory cells (e.g., T-cell memory) [47]; (iii) to produce the very first rapid response that may be crucial in a pandemic [48]; and (iv) to alter the immune response's length, precision, or affinities [49]. Refer Figure 4.3 to understand the reverse vaccinology utilizing the antigens.

4.4.2 Mechanisms of Adjuvant Action

In the formulation of vaccines, adjuvants are frequently used in order to enhance the success rate of vaccination. By enhancing the innate immune cells, they induce the vaccine's adaptive immunity [50]. The main goal of adjuvants is to enhance the development of antigen presenting signals and by activating APCs generate co-stimulatory signals. Major histocompatibility complexes (MHCs), which are antigen peptides that are shown on the surface of APCs following antigen absorption and processing, are known as antigen presentation signals. Co-stimulatory molecules (e.g., CD40, CD80, and CD86) expressed on the surface of APCs and released inflammatory cytokines [e.g., interleukin-6 (IL-6), IL-10, IL-12, and tumor necrosis factor-α (TNF-α)] are examples of co-stimulatory signals. The generation of these two signals has the potency to activate naive T cells, hence promoting an improved adaptive immune response [51]. APCs can produce signals 1 and 2 in response to immunostimulants such as damage-associated molecular patterns, pathogen-associated molecular patterns (PAMPs), and chemically produced small molecule agonists of TLRs. Delivery technologies that help deliver antigens on MHC molecules include lipid nanoparticles (LNPs), poly(lactide-co-glycolide) (PLGA), and caged protein nanoparticles. Notably, research has indicated that a variety of nanoparticle delivery methods are capable of specifically targeting B cells in order to maximize the effectiveness of an antibody response [52].

4.4.3 Types of Adjuvants

4.4.3.1 Aluminum Salts (Alum)

In human vaccine, aluminum salts are the most frequently used adjuvants, and they are also approved clinically. These adjuvants are mainly composed of aluminum hydroxide and phosphate precipitates used for vaccine antigen adsorption. Among the producers who prepare these formulations are

Brenntag Biosector, Chemtrade, and SPI PharmaTM. Alum is marketed under the trade names like Alhydrogel®, RehydragelTM, and Adju-Phos® [53]. The exact mechanism of action of alum is almost understood, and it is now evident that depot development is not necessary for alum to function as an adjuvant [54]. The main effect of alum as an adjuvant is innate immunity [55]. It promotes differentiation of B-cells to increase anti-body generation [56]. Additionally, it is widely known that alum incites mice's Th2 kind of immune response; nevertheless, practically, all vaccine antigens formulated with an alum adjuvant in humans cause a mixed type of response, i.e., Th1 and Th2 [57]. In *in vitro* study, it is found that, after priming macrophages and dendritic cells (DCs) with lipopolysaccharide (LPS), alum causes the NLRP3 inflammasome to produce IL-1β [58]. But *in vivo*, the observations are not supported by the adjuvanticity of alum [59]. It has been shown that alum induces the production of IL-4, IL-5, IgG1, and IgE and activates the Th2 type of immune response [60]. Alum also serves as a functional facilitator, employing DCs to carry out actin-mediated phagocytosis, which initiates a signaling cascade and activates two kinase proteins (Src and Syk). This, in turn, mobilizes Ca2+ and activates the transcription factor NFAT (calcineurin-nuclear factor of activated T cells), which, in turn, produces IL-2 [61]. Furthermore, alum exerts its adjuvanticity through the activation of the complement protein cascade. Because of its safety, ability to stabilize vaccination antigen, and ability to modulate long-lasting and high-production antibody titers, alum is a highly useful adjuvant. Vaccines prepared with an adjuvant such as alum cannot be frozen, lyophilized, or filter sterilized [62].

4.4.3.2 Toll-Like Receptor (TLR) Agonists

The innate immune response that is activated by vaccines formulated with adjuvants based on TLR ligands aids in the enhancement of the protective potential that the vaccine candidate imparts. MyD88 is an additional adaptor protein that is needed by TLR-based adjuvants in order to enhance the generation of antibodies by B cells and APCs. Furthermore, MyD88 signaling induces the development of germinal centers that are necessary for the production of antibody-secreting cells [63]. Two TLR ligand-dependent adjuvants that stimulate APCs properly are CpG DNA (TLR9-dependent adjuvant) and Porin B (TLR2-dependent adjuvant). They both are dependent on MyD88. PorB is a Neisseria meningitides outer membrane protein, and CpG is an unmethylated bacterial DNA pattern. TLR expression on APCs, such as macrophages, DCs, and B cells, has been shown. Consequently, the primary goal of utilizing a TLR agonist as an adjuvant in vaccine manufacturing is to activate these cells and induce a strong immunological response, thereby bridging the gap between adaptive immunity and innate immunity [63]. In addition to being employed in preclinical research, a number of TLR4 ligands have received Food and Drug Administration (FDA) approval for use in clinical trials such as MPLA and Bacillus Calmette-Guerin. In a mouse model, tumor regression was achieved by inducing acute inflammation *in vivo* using LPS, a well-known ligand for TLR4 [64]. It has been documented that, in a mouse model, intramuscular immunization with an enterotoxigenic E. coli vaccine antigen stimulates an increase in mucosal antibodies [65]. This is facilitated by the TLR4 agonist, a second-generation lipid adjuvant formulated in a stable emulsion (SLA-SE). Using the SARS-CoV-2 ferret model, a combination of TLR1/2 and TLR3 agonists was used to investigate the effectiveness of L-pampo. This resulted in a strong humoral and cell-mediated immune response. We believe that this development in adjuvant discovery for mucosal immunity can make a successful vaccine design possible and provide fresh hope for fighting pathogenic strains [66].

4.4.3.3 Saponins

Plant-derived saponins have attracted a lot of attention as potential adjuvants for vaccines. They have numerous biological and immunomodulatory characteristics [67]. Due to their extraordinary capacity to boost antibody responses and activate T helper type 1 cells (Th1) and cytotoxic T lymphocyte (CTL) immunological responses by stimulating DCs, Quil A and its pure fraction QS-21 are the most commonly utilized saponins as adjuvants. Furthermore, adjuvant systems (AS01, AS02, etc.) and immunostimulating complexes (ISCOMs) based on QS-21 have been created with lesser toxicity and better adjuvant effects [68]. AS01 has demonstrated good performance and safety when used as an adjuvant in marketed vaccines, such as the malaria RTS,S/AS01 vaccine (MosquirixTM) and the herpes zoster vaccine (Shingrix®) [69]. Adjuvants with license that contain saponins are now administered intramuscularly. Saponins are thought to have a great deal of potential to trigger robust mucosal and systemic immune responses. They have also been employed as adjuvants in nasal vaccinations in various investigations [70].

4.4.3.4 Liposomes and Lipid-Based Nanoparticles (LNPs)

The nanoscale lipid vesicles with nonviral and multifunctional properties are known as LNPs. They are made up of phospholipids, cholesterol, ionizable lipids, and polyethylene glycol modified lipids [71]. In LNPs, the major constituent is ionizable lipids. The integrity of structure is mainly contributed to the cholesterol and phospholipids. Polyethylene glycol modified lipids contribute to the LNPs stability. They produce their effect by three ways: firstly, by extending the bioavailability of antigens and effectively protecting their cargo; by enlarging the antigen's particle size to target APCs and encourage APC absorption; and by encouraging antigen escape through membrane fusion from endosomes, which triggers an immunological response from CD8+ T cells [72]. Following years of study, LNPs were demonstrated to significantly improve both humoral and cellular immune responses when used for giving a variety of vaccinations. Alameh and colleagues discovered that by stimulating the growth of potent T follicular helper cells, germinal center B cells, long-lived plasma cells, and memory B cells, LNPs improved humoral immune responses to various mRNA and protein subunit vaccines [73]. In order, to treat B16F10 melanoma in mice, Oberli and colleagues used LNPs to deliver mRNA vaccines for the tumor-associated antigens gp100 and TRP2. This resulted in tumor shrinkage and extended survival times for the mice. Additionally, the activation of CD8+ T cells was increased [74]. LNPs have been thoroughly investigated as a potential mRNA vaccine delivery method in recent years in clinical studies for vaccines against influenza, tumors, COVID-19, etc. In the year of 2020 BNT162b2 vaccine by Pfizer (also marketed as Comirnaty) and the mRNA-1273 vaccine by Moderna (marketed as Spikevax), both of which deliver mRNA antigens with LNPs were approved. The combat against COVID-19 has benefited greatly from the use of these two mRNA vaccines [75].

4.4.3.5 Polymer-Based Adjuvants

Natural and synthetic polymer particles are the two main categories in which the polymer particles are divided. Its mechanism of action is simple, and it is based on sustain release of antigens. Chitosan is a natural polymer that is widely used in vaccination. It has a higher cationic charge and better bio adhesive properties [76]. As a result, through electrostatic interaction, it can form strong complex with nucleic acid that is anionic in nature [77]. In addition, chitosan will remain in contact with surface of mucosa for a longer period of time for continuous immune cells stimulation by antigen due to its good bio adhesive properties. According to this, chitosan shows promise as a nucleic acid and mucosal vaccine delivery technology. Sawaengsak *et al.* created a nanoparticle vaccine by encapsulating inactivated influenza virus antigens in chitosan via sodium tripolyphosphate. Two doses of this vaccine were introduced into mice intraperitoneally and observed that mice were able to develop a greater number of IFNγ+ T cells and antibodies specific to the antigen. The mice who received the vaccination exhibited complete defence against infection caused by influenza virus [78]. Chitosan has recently undergone testing as a Neocrown DNA vaccine delivery method. High concentrations of neutralizing antibodies were found in mice following intranasal immunization, and several coronaviruses as well as their mutant pseudoviruses were successfully neutralized [79]. Furthermore, it has recently been discovered that chitosan activates the DNA sensor cyclic GMP-AMP synthase-stimulator of interferon genes (cGAS-STING) pathway, which, in turn, induces the production of type I interferons for CD8+ T-cell immunity [80]. Compared to natural polymers, synthetic polymer particles usually exhibit greater repeatability and a more regulated rate of gradual release [81]. Nowadays, PLGA, a synthetic polymeric particle material, is frequently utilized as a vaccine delivery carrier. In order to improve antigen uptake and presentation by DCs, Kim *et al.* conducted preclinical trials wherein they encapsulated vaccine components (i.e., antigens and immunostimulants) in polylactic acid (PLGA). This allowed the immunostimulant to target DCs in lymph nodes rather than systemic spread [82]. A PLGA particle containing antigens and double-stranded ribonucleic acid (dsRNA) adjuvants was created by Koerner *et al.* It was discovered that the particle vaccine had a perfect release curve, expanded the number of targeted lymph nodes, and was efficiently phagocytosed and presented by DCs. Ultimately, an efficient and long-lasting immune response against cancer was generated [83]. Additionally, PLGA is employed in the delivery of HBV and other cancer vaccines, as well as in enhancing their immunotherapeutic effects [84]. PLGA particles are currently undergoing clinical trials to be used as cancer vaccine delivery systems (NCT01753089 and NCT04751786) [85]. It is noteworthy that the FDA has given PLGA approval for the delivery of drugs. Thus, we think that a vaccine delivery system based on PLGA will soon be authorized for use in humans [86].

4.5 Novel Formulation Strategies for Improved Vaccine Efficacy

4.5.1 Biological Adjuvants

Biological adjuvants are a diverse group of compounds derived from living organisms or their products. They enhance the immune response to antigens, unlike synthetic adjuvants, which are chemically synthesized. Microbial adjuvants come from bacteria, viruses, or fungi. Examples include bacterial cell wall components like LPS and viral proteins. Certain cytokines and chemokines regulate immune responses. Adjuvants based on these molecules can modulate the immune system by promoting immune cell activation [87]. Some toxins stimulate immune cells or induce inflammation, making them potent adjuvants. Cholera toxin and its derivatives have been explored as mucosal adjuvants. DNA and RNA can act as adjuvants by activating innate immune receptors like TLRs. Synthetic oligonucleotides containing CpG motifs mimic bacterial DNA and boost immune responses. Extracts from plants or algae can exhibit adjuvant properties by interacting with immune cells or modulating cytokine production [88].

They activate innate immune receptors, leading to the production of pro-inflammatory molecules that enhance antigen presentation and immune cell activation. Adjuvants promote the maturation of DCs and other APCs, increasing antigen uptake and presentation to T cells. They upregulate co-stimulatory molecules on APCs, crucial for T-cell activation and differentiation. Some adjuvants induce cytokine production that favors specific immune responses, such as Th1 cytokines for cellular immunity or Th2 cytokines for humoral immunity [89].

A study on a candidate COVID-19 vaccine, Ad26.COV2.S, used a multicenter, placebo-controlled trial to evaluate its safety and effectiveness. Participants aged 18–55 and those 65 and older received varying doses of the vaccine or a placebo. The main outcomes assessed were adverse events and immune response. Following vaccine administration, common adverse events included fatigue, headache, myalgia, and injection-site pain. Systemic adverse events like fever were less frequent in older participants and those who received lower vaccine doses. Reactogenicity was lower after the second dose [90]. Most participants developed neutralizing antibodies against the virus, with higher titers after the second dose. Antibody responses were similar across age groups. Additionally, T-cell responses were detected in a majority of participants, indicating a robust immune reaction. The study concludes that Ad26.COV2.S shows promise in terms of safety and immune response, supporting its further development as a COVID-19 vaccine candidate [91]. After receiving varying doses of the VSV-Ebola vaccine, researchers observed differences in the human antibody response among healthy adult volunteers. Those who received 20 million plaque-forming units (PFU) showed a greater diversity of antibody epitopes related to the Ebola glycoprotein (GP) compared to those who received 3 million or 100 million PFU. Antibodies binding to GP were significantly higher in individuals vaccinated with 20 million or 100 million PFU, correlating strongly with their ability to neutralize the virus. However, a second vaccination did not boost antibody levels or virus neutralization, and there was minimal improvement in antibody affinity. IgM antibodies remained predominant even after the second vaccination and contributed significantly to virus neutralization in laboratory tests. These findings have implications for identifying new vaccine targets and improving countermeasures against Ebola [92].

Virus-like particles (VLPs) have become a focal point in virology and vaccine development due to their unique structure and immunogenic properties. VLPs are complex structures resembling viruses but without genetic material. They consist of viral proteins that self-assemble into virus-like shapes, ensuring safety without the risk of viral replication. Upon administration, VLPs are recognized by immune cells like DCs and macrophages. The orderly presentation of viral antigens on VLP surfaces facilitates recognition by immune receptors, initiating crucial immune responses. VLPs bind to immune receptors on APCs, activating innate immune pathways. This activation leads to the production of key cytokines and signaling molecules essential for immune response coordination [93]. VLPs promote the maturation of DCs, which then migrate to lymphoid organs to present VLP-derived antigens to T cells. This process enhances antigen presentation and primes T cells for activation. VLP-induced maturation of DCs triggers adaptive immune responses. T helper cells recognize VLP antigens, leading to the activation of specific immune pathways

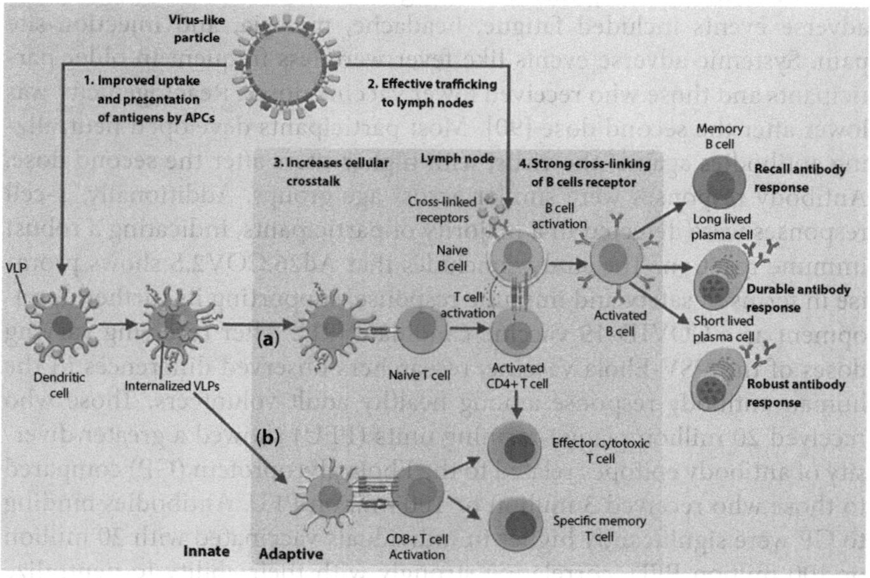

Figure 4.4 Induction of both innate and adaptive immunity via VLP. Adopted under CC BY 4 from [95]. Adopted (2021), Front. Microbiol.

tailored to combat the target virus or antigen. VLPs activate B cells, leading to antibody production and affinity maturation. This results in the generation of high-quality antibodies crucial for immune defense. VLP vaccination establishes long-lasting immune memory, providing rapid and effective protection upon re-exposure to the target virus. This memory is vital for sustained immunity and vaccine efficacy. VLPs can be engineered to enhance immunogenicity or modulate immune responses for specific therapeutic goals, making them versatile tools in vaccine design and immunotherapy [94]. Refer to Figure 4.4 for the pictorial presentation.

4.5.2 Biodegradable Polymers

Biodegradable polymers have garnered significant interest across various domains, notably in medicine and vaccine development. Within the vaccine context, adjuvants hold pivotal roles in augmenting immune responses and bolstering vaccine effectiveness. Biodegradable polymers stand out as promising adjuvants due to their capacity to modulate immune reactions, sustain antigen release, and reduce the necessity for multiple vaccine doses. Polylactic acid (PLA), polyglycolic acid (PGA), and their copolymers, like PLGA, are extensively utilized biodegradable polymers in vaccine adjuvant formulations. Renowned for their biocompatibility, these polymers can be customized to achieve specific release profiles [96]. Another notable polymer is polycaprolactone (PCL), prized for its slow degradation rate, enabling prolonged antigen presentation and immune activation. Chitosan, derived from chitin, has also emerged as a compelling adjuvant due to its mucoadhesive properties and immunomodulatory potential. Biodegradable polymers excel at encapsulating antigens, safeguarding them from degradation and facilitating uptake by APCs like DCs. Their ability to release antigens in a controlled manner ensures sustained exposure, fostering prolonged immune activation and memory cell formation. Biodegradable polymers can modulate immune responses by steering Th1 or Th2 pathways, bolstering antibody production, or triggering cellular immunity, contingent upon their properties and formulation [97].

A strategy was developed to improve targeted delivery of antigens using polyanhydride nanoparticles. This involved decorating the nanoparticle surface with specific carbohydrates, mimicking features found on respiratory pathogens. By targeting specific receptors on alveolar macrophages (AMϕ), key immune cells in the respiratory tract, the nanoparticles were designed to enhance immune responses. The surface of polyanhydride nanoparticles was modified with galactose and di-mannose, two common carbohydrate structures found on respiratory pathogens. This modification aimed to interact with CLR receptors on AMϕ. When compared to non-functionalized nanoparticles, the decorated nanoparticles significantly increased the expression of key immune markers on AMϕ, including MHC I and II, CD86, CD40, and C-type lectin receptors (CLR) CIRE. Furthermore, the functionalization with di-mannose and galactose increased the expression of specific receptors on AMϕ, such as the macrophage mannose receptor (MMR) and the macrophage galactose lectin. These changes in receptor expression correlated with enhanced activation of AMϕ, as indicated by increased production of pro-inflammatory cytokines like IL-1β, IL-6, and TNF-α. Importantly, the enhanced activation and uptake of functionalized nanoparticles by AMϕ were found to depend on the presence of the MMR receptor. This targeted engagement of MMR and other CLR receptors represents a promising strategy to boost the adjuvant properties of nanovaccine formulations, particularly for promoting strong immune responses in the lungs [98].

4.5.3 Designer Adjuvants with Specific Immunomodulatory Properties

Immunomodulation involves regulating immune responses to achieve desired outcomes. Designer adjuvants are designed to modulate specific aspects of immunity while minimizing undesirable effects, such as inflammation or autoimmunity. Specific immunomodulatory properties in designer adjuvants are essential for targeted immune activation, reducing side effects, and tailoring vaccine responses to different populations. Designer adjuvants are developed using molecular design strategies, including structural modifications, incorporation of immunomodulatory molecules, and nanoparticle-based delivery systems. Examples include TLR agonist-based adjuvants, cytokine-modified adjuvants, and peptide-based adjuvants, each targeting specific immune pathways for tailored immune responses. Designer adjuvants are crucial for infectious disease vaccines, cancer immunotherapies, and autoimmune disease therapeutics, where precise immune modulation is critical for efficacy [99].

Toll-like receptors (TLRs) form a crucial part of the immune system by recognizing specific patterns on pathogens called PAMPs. This recognition triggers a series of events aimed at mounting an effective immune response. Adjuvants that target TLRs are designed to mimic the presence of pathogens, thereby enhancing the immune response to vaccines. They work by activating TLRs on immune cells, leading to the production of inflammatory cytokines, increased antigen presentation, and activation of immune cells like DCs and T cells. TLR agonists bind to TLRs on immune cells, initiating signaling pathways that activate transcription factors. This activation leads to the production of pro-inflammatory cytokines crucial for immune responses. TLR activation enhances antigen presentation by APCs, aiding in T-cell activation. TLR agonists increase the expression of molecules important for T-cell activation [100]. These effects result in an enhanced adaptive immune response, including increased antibody production and cytotoxic T-cell activity. Different types of TLR agonists have been studied for their adjuvant properties, including those targeting TLR-4 (e.g., LPS and MPLA), TLR-3 (e.g., Poly I:C), TLR-7/8 (e.g., Imiquimod and resiquimod), and TLR-9 (e.g., CpG ODNs) [101].

4.5.4 Adjuvanted Mucosal Vaccines

Mucosal surfaces are protected by specialized epithelial cells and mucosal-associated lymphoid tissues, such as Peyer's patches in the gut, tonsils in the oral cavity, and bronchial-associated lymphoid tissue in the respiratory tract. These areas are rich in immune cells like T cells, B cells, DCs, and microfold (M) cells that specialize in capturing antigens. The immune defense at mucosal sites is crucial for preventing infections by pathogens that enter through these surfaces. It involves various mechanisms such as producing secretory immunoglobulin A (sIgA), secreting mucin, and activating innate immune cells [102]. Mucosal vaccines are designed to strengthen these protective mechanisms to confer immunity against pathogens. Adjuvants help M cells and DCs take up antigens more effectively, improving their presentation to immune cells. They activate innate immune receptors like TLRs on mucosal epithelial cells and immune cells, triggering local immune responses. Adjuvants aid in inducing mucosal immune responses, including sIgA production and recruiting effector cells to mucosal sites. Mucosal immunostimulants, such as cholera toxin (CT) and heat-labile enterotoxin (LT), activate mucosal immune cells and improve antigen uptake, although modifications are needed to prevent toxicity. Nanoparticles, liposomes, and VLPs act as delivery systems, enhancing antigen stability and presentation to immune cells. Synthetic molecules like CpG oligodeoxynucleotides (ODNs) targeting TLR9 and LPS derivatives mimic pathogen patterns, boosting immune responses. Recombinant proteins or peptides like interleukins (e.g., IL-12) and chemokines (e.g., CCL19) modulate immune cell activities, aiding in immune cell recruitment and activation [103].

Nasal-associated lymphoid tissues (NALTs) are specialized lymphoid organs located in the submucosa of the nasal passage. They play a crucial role in the immune response to inhaled antigens. However, the mechanisms governing immunity induction within NALTs are not fully understood. Research has shown that under normal conditions, conventional DCs (cDCs) in NALTs suppress T-cell responses. These cDCs, also found in human NALTs such as tonsils and adenoids, exhibit a unique gene expression pattern and inhibit T-cell proliferation through mechanisms independent of direct contact. This suppression can be reduced by blocking reactive oxygen species and prostaglandin E2. During localized virus infections or inflammation, monocyte-derived DCs (moDCs) are recruited to NALTs, altering the composition and function of DCs in this region. This recruitment dilutes the suppressive cDC population and allows for local T-cell activation and priming. To leverage these insights, researchers developed an intranasal vaccine delivery system. This system aims to recruit moDCs while ensuring sustained antigen release into NALTs. By doing so, they observed significantly improved T-cell responses following intranasal immunization [104].

4.6 Future Directions and Challenges

4.6.1 Personalized Vaccines and Adjuvant Selection

Immune responses to personalized vaccines are based on complicated interactions of host genetic and environmental factors. The development of personalized vaccines is a multistep process. Understanding the correlations between an immune response gene polymorphism and variations in vaccine immune responses is important. This understanding is useful in designing a vaccine or adjuvant. Hepatitis B vaccine immune response was restricted due to insertion/deletion of variants at the IL-12B locus and HLA-DRB1*07 and cytokine SNPs at the IL-2 and IL-4 loci. A new HBV vaccine that consists of a peptide as novel epitopes (identified from chronic carriers) and cytokine adjuvants (for example, GM-CSF) found to circumvent hepatitis B vaccine non-response. SNPs negatively impact the ability of measles vaccine virus to bind to its receptor and, hence, prevent induction of immunity. Six deoxynucleotide-long DNA (adjuvant) sequences with a monophosphoryl lipid-A or a central CpG dinucleotide activate TLRs (TLR9 or TLR4, respectively) and, thus, circumvent the restrictions in binding of measles vaccine virus to other receptors [105, 106].

4.6.2 Novel Adjuvant Discovery Platforms

Technological and computational advancements have revolutionized adjuvant discovery, enabling researchers to gather vast amounts of data for mechanistic understanding and efficacy evaluation. Next-generation sequencing, omics approaches, machine learning, processing yield hidden patterns and features, and high-content imaging can collect system-level data, enabling *in silico* binding simulations and transforming the adjuvant development workflow. These approaches can be implemented at all stages of adjuvant design. For adjuvant discovery, *in silico* experimentation is frequently combined with machine learning, an artificial intelligence technology, to create algorithmic models for probabilistic predictions. Active learning, reinforcement learning, and Bayesian optimization are three popular black-box algorithms. High-throughput adjuvant screening employs Bayesian optimization, a Gaussian process regression that predicts expected attributes. High-content imaging used in adjuvant discovery visualizes and quantifies therapeutic interactions in cell populations employing multiparameter algorithms and fluorescent labeling [107].

4.6.3 Addressing Safety Concerns of New Adjuvants

Safety concerns in the development of new adjuvants are important for their suitability for use in vaccines. Strategies include preclinical safety assessment, structural characterization, dose optimization, administration route, formulation optimization, adjuvant–antigen interaction studies, immunotoxicity assessment, long-term safety monitoring, and regulatory oversight. Preclinical studies evaluate toxicity, pharmacokinetics, biodistribution, and potential adverse effects in animal models. Structural characterization techniques provide insights into adjuvant composition, stability, and interactions/reactogenicity with biological molecules. Dose optimization determines the optimal dose for achieving the desired immune response without causing undue toxicity. The route of administration can also influence the safety profile. Long-term safety monitoring ensures continuous assessment of benefit-risk profile. Regulatory oversight ensures adjuvants adhere to established safety standards [108, 109].

4.7 Conclusion

Novel formulations for vaccines offer numerous benefits, including improved stability, enhanced immunity, targeted delivery, dose-sparing strategies, combination vaccines, mucosal vaccination, painless administration, and customization. These formulations can withstand various storage conditions without refrigeration, making them ideal for regions with limited cold chain infrastructure. They also boost the immune response, improving efficacy in populations with weak or compromised immune systems. Targeted delivery systems, such as nanoparticles or liposomes, reduce off-target effects and enhance the safety profile of vaccines. Dose-sparing strategies can stretch limited vaccine supplies, increase availability, and reduce costs during vaccine shortages or outbreaks. Novel formulations also facilitate mucosal vaccination, promoting mucosal immunity and preventing infections. These formulations also improve patient adherence and acceptance, simplifying vaccine delivery in resource-limited settings.

References

1. Bhardwaj, S., Vaccines, in: *Pharmaceutical Medicine and Translational Clinical Research*, D. Vohora and G. Singh (Eds.), pp. 341–353, Academic Press Inc, 2018.
2. Segura-Collar, B., Hiller-Vallina, S., de Dios, O., Caamaño-Moreno, M., Mondejar-Ruescas, L., Sepulveda-Sanchez, J.M., Gargini, R., Advanced immunotherapies for glioblastoma: tumor neoantigen vaccines in combination with immunomodulators. *Acta Neuropathol. Commun.*, 11, 1, 79, 2023.
3. Szwejser-Zawislak, E., Wilk, M.M., Piszczek, P., Krawczyk, J., Wilczyńska, D., Hozbor, D., Evaluation of whole-cell and acellular pertussis vaccines in the context of long-term herd immunity. *Vaccines (Basel)*, 11, 1, 1, 2022.
4. Kumar, S., Live-attenuated bacterial vectors for delivery of mucosal vaccines, DNA vaccines, and cancer immunotherapy. *Pharm. Microbes*, 26, 39–73, 2019.
5. Pöyhönen, L., Bustamante, J., Casanova, J.L., Jouanguy, E., Zhang, Q., Life-threatening infections due to live-attenuated vaccines: Early manifestations of inborn errors of immunity. *J. Clin. Immunol.*, 39, 4, 376–390, 2019.
6. Ellebedy, A.H. and Ahmed, R., Antiviral vaccines: Challenges and advances, in: *The Vaccine Book*, pp. 283–310, 2016.
7. Cid, R. and Bolívar, J., Platforms for production of protein-based vaccines: From classical to next-generation strategies. *Biomolecules*, 11, 8, 1072, 2021.
8. Nooraei, S., Bahrulolum, H., Hoseini, Z.S., Katalani, C., Hajizade, A., Easton, A.J., Ahmadian, G., Virus-like particles: preparation, immunogenicity and their roles as nanovaccines and drug nanocarriers. *J. Nanobiotechnol.*, 19, 1, 59, 2021.
9. Moradi Vahdat, M., Hemmati, F., Ghorbani, A., Rutkowska, D., Afsharifar, A., Eskandari, M.H., Rezaei, N., Niazi, A., Hepatitis B core-based virus-like

particles: A platform for vaccine development in plants. *Biotechnol. Rep. (Amst.)*, 29, e00605, 2021.
10. Nascimento, I.P. and Leite, L.C., Recombinant vaccines and the development of new vaccine strategies. *Braz. J. Med. Biol. Res.*, 45, 12, 1102–1011, 2012.
11. Yao, Y., Zhang, Z., Yang, Z., The combination of vaccines and adjuvants to prevent the occurrence of high incidence of infectious diseases in bovine. *Front. Vet. Sci.*, 10, 1243835, 2023.
12. Conjugate vaccine, in: *Australian Immunization Handbook*, Available from https://immunisationhandbook.health.gov.au/technical-terms/conjugate-vaccine. Accessed on 18 March, 2024.
13. Goldblatt, D., Conjugate vaccines. *Clin. Exp. Immunol.*, 119, 1–3, 2000.
14. Danino, D., van der Beek, B.A., Greenberg, D., Ben-Shimol, S., Dagan, R., The impact of pneumococcal conjugate vaccine on ceftriaxone consumption in the community among young children. *Int. J. Infect. Dis.*, 135, 21–27, 2023.
15. Kutzler, M.A. and Weiner, D.B., DNA vaccines: ready for prime time? *Nat. Rev. Genet.*, 9, 10, 776–788, 2008.
16. Ramakrishna, L., Anand, K.K., Mohankumar, K.M., Ranga, U., Codon optimization of the tat antigen of human immunodeficiency virus type 1 generates strong immune responses in mice following genetic immunization. *J. Virol.*, 78, 17, 9174–9189, 2004.
17. Frelin, L., Ahlén, G., Alheim, M., Weiland, O., Barnfield, C., Liljeström, P., Sällberg, M., Codon optimization and mRNA amplification effectively enhances the immunogenicity of the hepatitis C virus nonstructural 3/4A gene. *Gene Ther.*, 11, 6, 522–533, 2004.
18. Khalid, K. and Poh, C.L., The development of DNA vaccines against SARS-CoV-2. *Adv. Med. Sci.*, 68, 2, 213–226, 2023.
19. Dai, X., Xiong, Y., Li, N., Jian, C., Vaccine types, in: *Vaccines – the history and future*, V. Kumar (Ed.), InTech Open, UK, 2019.
20. Dean, L., Blood group antigens are surface markers on the red blood cell membrane, in: *Blood groups and red cell antigens*, National Center for Biotechnology Information, Bethesda, MD, USA, 2005. Available from: https://www.ncbi.nlm.nih.gov/books/NBK2264/.
21. He, Y., Liu, X., Huang, Y., Zou, Z., Chen, H., Lai, H., Zhang, L., Wu, Q., Zhang, J., Wang, S., Zhang, J., Tao, A., Sun, B., Reduction of the number of major representative allergens: from clinical testing to 3-dimensional structures. *Mediators Inflamm.*, 2014, 291618, 2014.
22. Zhang, J. and Tao, A., Antigenicity, immunogenicity, allergenicity. *Allergy Bioinf.*, 8, 175–186, 2015.
23. Hagel, I., Cabrera, M., Puccio, F., Santaella, C., Buvat, E., Infante, B., Zabala, M., Cordero, R., Di Prisco, M., Co-infection with *Ascaris lumbricoides* modulates protective immune responses against *Giardia duodenalis* in school Venezuelan rural children. *Acta Trop.*, 117, 3, 189–195, 2011.

24. Palm, N., Rosenstein, R., Medzhitov, R., Allergic host defences. *Nature*, 484, 7395, 465–472, 2012.
25. Sepp, E., Julge, K., Vasar, M., Naaber, P., Bjorksten, B., Mikelsaar, M., Intestinal microflora of Estonian and Swedish infants. *Acta Paediatr.*, 86, 9, 956–961, 1997.
26. Hesselmar, B., Aberg, N., Aberg, B., Eriksson, B., Björkstén, B., Does early exposure to cat or dog protect against later allergy development? *Clin. Exp. Allergy*, 29, 5, 611–617, 2008.
27. Liu, A.H. and Murphy, J.R., Hygiene hypothesis: fact or fiction? *J. Allergy Clin. Immunol.*, 111, 3, 471–478, 2003.
28. Maizels, R.M., McSorley, H.J., Smyth, D.J., Helminths in the hygiene hypothesis: sooner or later? *Clin. Exp. Immunol.*, 177, 1, 38–46, 2014.
29. Majak, P., Brzozowska, A., Bobrowska-Korzeniowska, M., Stelmach, I., Early exposure to unhygienic conditions and infections is associated with expression of different Toll-like receptors. *J. Investig. Allergol. Clin. Immunol.*, 19, 4, 260–265, 2009.
30. Michel, S., Busato, F., Genuneit, J., Pekkanen, J., Dalphin, J.C., Riedler, J., Mazaleyrat, N., Weber, J., Karvonen, A.M., Hirvonen, M.R., Braun-Fahrländer, C., Lauener, R., von Mutius, E., Kabesch, M., Tost, J., Pasture study group. Farm exposure and time trends in early childhood may influence DNA methylation in genes related to asthma and allergy. *Allergy*, 68, 3, 355–364, 2013.
31. Min, K.B. and Min, J.Y., Exposure to household endotoxin and total and allergen-specific IgE in the US population. *Environ. Pollut.*, 199, 148–154, 2015.
32. Palmer, L.J., Celedon, J.C., Weiss, S.T., Wang, B., Fang, Z., Xu, X., *Ascaris lumbricoides* infection is associated with increased risk of childhood asthma and atopy in rural China. *Am. J. Respir. Crit. Care Med.*, 165, 11, 1489–1493, 2002.
33. Ghosh, U., Sayef Ahammed, K., Mishra, S., Bhaumik, A., The emerging roles of silver nanoparticles to target viral life cycle and detect viral pathogens. *Chem. Asian J.*, 17, 5, e202101149, 2022.
34. Murphy, F.A., Historical perspective: What constitutes discovery (of a new virus)? *Adv. Virus Res.*, 95, 197–220, 2016.
35. Moelling, K., Viruses more friends than foes. *Electroanalysis*, 32, 4, 669–673, 2020.
36. Lauring, A.S., Frydman, J., Andino, R., The role of mutational robustness in RNA virus evolution. *Nat. Rev. Microbiol.*, 11, 5, 327–336, 2013.
37. Geoghegan, J.L., Duchêne, S., Holmes, E.C., Comparative analysis estimates the relative frequencies of co-divergence and cross-species transmission within viral families. *PLoS Pathog.*, 13, 2, e1006215, 2017.
38. Elechiguerra, J.L., Burt, J.L., Morones, J.R., Camacho-Bragado, A., Gao, X., Lara, H.H., Yacaman, M.J., Interaction of silver nanoparticles with HIV-1. *J. Nanobiotechnol.*, 3, 6, Jun 29.

39. Gaikwad, S., Ingle, A., Gade, A., Rai, M., Falanga, A., Incoronato, N., Russo, L., Galdiero, S., Galdiero, M., Antiviral activity of mycosynthesized silver nanoparticles against herpes simplex virus and human parainfluenza virus type 3. *Int. J. Nanomed.*, 8, 4303–4314, 2013.
40. Wagner, L., Verma, A., Meade, B.D., Reiter, K., Narum, D.L., Brady, R.A., Little, S.F., Burns, D.L., Structural and immunological analysis of anthrax recombinant protective antigen adsorbed to aluminum hydroxide adjuvant. *Clin. Vaccine Immunol.*, 19, 9, 1465–1473, 2012.
41. Nick, P.C., Scholtz, J.M., Grimsley, G.R., Forces stabilizing proteins. *FEBS Lett.*, 588, 14, 2177–2184, 2014.
42. Scheiblhofer, S., Laimer, J., Machado, Y., Weiss, R., Thalhamer, J., Influence of protein fold stability on immunogenicity and its implications for vaccine design. *Expert Rev. Vaccines*, 16, 5, 479–489, 2017.
43. Swoboda, I., Bugajska-Schretter, A., Linhart, B., Verdino, P., Keller, W., Schulmeister, U., Sperr, W.R., Valent, P., Peltre, G., Quirce, S., Douladiris, N., Papadopoulos, N.G., Valenta, R., Spitzauer, S., A recombinant hypoallergenic parvalbumin mutant for immunotherapy of IgE-mediated fish allergy. *J. Immunol.*, 178, 10, 6290–6296, 2007.
44. Reed, S.G., Orr, M.T., Fox, C.B., Key roles of adjuvants in modern vaccines. *Nat. Med.*, 19, 12, 1597–1608, 2013.
45. Fox, C.B., Kramer, R.M., Barnes, V.L., Dowling, Q.M., Vedvick, T.S., Working together: interactions between vaccine antigens and adjuvants. *Ther. Adv. Vaccines*, 1, 1, 7–20, 2013.
46. Fagain, C.O., Understanding and increasing protein stability. *Biochim. Biophys. Acta*, 1252, 1, 1–14, 1995.
47. Saylor, K., Gillam, F., Lohneis, T., Zhang, C., Designs of antigen structure and composition for improved protein-based vaccine efficacy. *Front. Immunol.*, 11, 283, 2020.
48. Ward, A.B. and Wilson, I.A., Innovations in structure-based antigen design and immune monitoring for next generation vaccines. *Curr. Opin. Immunol.*, 65, 50–56, 2020.
49. Dormitzer, P.R., Ulmer, J.B., Rappuoli, R., Structure-based antigen design: a strategy for next generation vaccines. *Trends Biotechnol.*, 26, 12, 659–667, 2008.
50. Veggiani, G., Giabbai, B., Semrau, M.S., Medagli, B., Riccio, V., Bajc, G., Storici, P., de Marco, A., Comparative analysis of fusion tags used to functionalize recombinant antibodies. *Protein Expr. Purif.*, 166, 105505, 2020.
51. Basu, A., Li, X., Leong, S.S.J., Refolding of proteins from inclusion bodies: Rational design and recipes. *Appl. Microbiol. Biotechnol.*, 92, 241–251, 2011.
52. Deyev, S.M. and Lebedenko, E.N., Multivalency: The hallmark of antibodies used for optimization of tumor targeting by design. *BioEssays*, 30, 9, 904–918, 2008.
53. Tong, J.C., Tan, T.W., Ranganathan, S., Methods and protocols for prediction of immunogenic epitopes. *Brief Bioinform.*, 8, 2, 96–108, 2007.

54. Theiler, J. and Korber, B., Graph-based optimization of epitope coverage for vaccine antigen design, in: *Statistics in Medicine*, pp. 181–194, John Wiley and Sons Ltd, 2018.
55. Uray, K. and Hudecz, F., Peptide epitopes: Identification and structural modifications of synthetic antigens. *Amino Acids, Pept. Proteins*, 39, 68–113, 2015.
56. Rappuoli, R., Bottomley, M.J., D'Oro, U., Finco, O., De Gregorio, E., Reverse vaccinology 2.0: Human immunology instructs vaccine antigen design. *J. Exper. Med.*, 213, 4, 469–481, 2016.
57. Burton, D.R., What Are the Most Powerful Immunogen Design Vaccine Strategies? Reverse Vaccinology 2.0 Shows Great Promise. *Cold Spring Harb Perspect Biol.*, 9, 11, a030262, 2017.
58. Li, H., Nookala, S., Re, F., Aluminum Hydroxide Adjuvants Activate Caspase-1 and Induce IL-1β and IL-18 Release. *J. Immunol.*, 178, 8, 5271–5276, 2007.
59. Marrack, P., McKee, A.S., Munks, M.W., Towards an understanding of the adjuvant action of aluminium. *Nat. Rev. Immunol.*, 9, 4, 287–293, 2009.
60. Lindblad, E.B., Aluminium adjuvants—in retrospect and prospect. *Vaccine*, 22, 27–28, 3658–3668, 2004.
61. Ong, G.H., Lian, B.S.X., Kawasaki, T., Kawai, T., Exploration of Pattern Recognition Receptor Agonists as Candidate Adjuvants. *Front. Cell. Infect. Microbiol.*, 11, 745016, 2021.
62. Gupta, R., Aluminum compounds as vaccine adjuvants. *Adv. Drug Deliv. Rev.*, 32, 3, 155–172, 1998.
63. Mosaheb, M.M., Reiser, M.L., Wetzler, L.M., Toll-Like Receptor Ligand-Based Vaccine Adjuvants Require Intact MyD88 Signaling in Antigen-Presenting Cells for Germinal Center Formation and Antibody Production. *Front. Immunol.*, 8, 2017. [Internet] Available from: http://journal.frontiersin.org/article/10.3389/fimmu.2017.00225/full.
64. Shetab Boushehri, M.A. and Lamprecht, A., TLR4-Based Immunotherapeutics in Cancer: A Review of the Achievements and Shortcomings. *Mol. Pharm.*, 15, 11, 4777–4800, 2018.
65. Liang, H., Poncet, D., Seydoux, E. *et al.*, The TLR4 agonist adjuvant SLA-SE promotes functional mucosal antibodies against a parenterally delivered ETEC vaccine. *NPJ Vaccines*, 4, 1, 19, 2019.
66. Jeong, S.-K., Heo, Y.-K., Jeong, J.-H. *et al.*, COVID-19 Subunit Vaccine with a Combination of TLR1/2 and TLR3 Agonists Induces Robust and Protective Immunity. *Vaccines*, 9, 9, 957, 2021.
67. Sparg, S.G., Light, M.E., Van Staden, J., Biological activities and distribution of plant saponins. *J. Ethnopharmacol.*, 94, 2–3, 219–243, 2004.
68. Marciani, D.J., Elucidating the Mechanisms of Action of Saponin-Derived Adjuvants. *Trends Pharmacol. Sci.*, 39, 6, 573–585, 2018.
69. Lacaille-Dubois, M.-A., Updated insights into the mechanism of action and clinical profile of the immunoadjuvant QS-21: A review. *Phytomedicine*, 60, 152905, 2019.

70. Hu, K.-F., Lövgren-Bengtsson, K., Morein, B., Immunostimulating complexes (ISCOMs) for nasal vaccination. *Adv. Drug Deliv. Rev.*, 51, 1–3, 149–159, 2001.
71. Oberli, M.A., Reichmuth, A.M., Dorkin, J.R. *et al.*, Lipid Nanoparticle Assisted mRNA Delivery for Potent Cancer Immunotherapy. *Nano Lett.*, 17, 3, 1326–1335, 2017.
72. Samaridou, E., Heyes, J., Lutwyche, P., Lipid nanoparticles for nucleic acid delivery: Current perspectives. *Adv. Drug Deliv. Rev.*, 154–155, 37–63, 2020.
73. Alameh, M.-G., Tombácz, I., Bettini, E. *et al.*, Lipid nanoparticles enhance the efficacy of mRNA and protein subunit vaccines by inducing robust T follicular helper cell and humoral responses. *Immunity*, 55, 6, 1136–1138, 2022.
74. Oberli, M.A., Reichmuth, A.M., Dorkin, J.R. *et al.*, Lipid Nanoparticle Assisted mRNA Delivery for Potent Cancer Immunotherapy. *Nano Lett.*, 17, 3, 1326–1335, 2017.
75. Polack, F.P., Thomas, S.J., Kitchin, N. *et al.*, Safety and Efficacy of the BNT162b2 mRNA Covid-19 Vaccine. *N. Engl. J. Med.*, 383, 27, 2603–2615, 2020.
76. Wu, D., Zhu, L., Li, Y. *et al.*, Chitosan-based Colloidal Polyelectrolyte Complexes for Drug Delivery: A Review. *Carbohydr. Polym.*, 238, 116126, 2020.
77. Kumar, M., Behera, A.K., Lockey, R.F. *et al.*, Intranasal Gene Transfer by Chitosan–DNA Nanospheres Protects BALB/c Mice Against Acute Respiratory Syncytial Virus Infection. *Hum. Gene Ther.*, 13, 12, 1415–1425, 2002.
78. Sawaengsak, C., Mori, Y., Yamanishi, K., Mitrevej, A., Sinchaipanid, N., Chitosan Nanoparticle Encapsulated Hemagglutinin-Split Influenza Virus Mucosal Vaccine. *AAPS PharmSciTech.*, 15, 2, 317–325, 2014.
79. Kumar, U.S., Afjei, R., Ferrara, K., Massoud, T.F., Paulmurugan, R., Gold-Nanostar-Chitosan-Mediated Delivery of SARS-CoV-2 DNA Vaccine for Respiratory Mucosal Immunization: Development and Proof-of-Principle. *ACS Nano*, 15, 11, 17582–17601, 2021.
80. Carroll Elizabeth, C., Jin, L., Mori, A. *et al.*, The Vaccine Adjuvant Chitosan Promotes Cellular Immunity via DNA Sensor cGAS-STING-Dependent Induction of Type I Interferons. *Immunity*, 44, 3, 597–608, 2016.
81. Karlsson, J., Vaughan, H.J., Green, J.J., Biodegradable Polymeric Nanoparticles for Therapeutic Cancer Treatments. *Annu. Rev. Chem. Biomol. Eng.*, 9, 1, 105–127, 2018.
82. Kim, H., Niu, L., Larson, P. *et al.*, Polymeric nanoparticles encapsulating novel TLR7/8 agonists as immunostimulatory adjuvants for enhanced cancer immunotherapy. *Biomaterials*, 164, 38–53, 2018.
83. Koerner, J., Horvath, D., Herrmann, V.L. *et al.*, PLGA-particle vaccine carrying TLR3/RIG-I ligand Riboxxim synergizes with immune checkpoint blockade for effective anti-cancer immunotherapy. *Nat. Commun.*, 12, 1, 2935, 2021.

84. Xie, X., Hu, Y., Ye, T. *et al.*, Therapeutic vaccination against leukaemia via the sustained release of co-encapsulated anti-PD-1 and a leukaemia-associated antigen. *Nat. Biomed. Eng.*, 5, 5, 414–428, 2020.
85. Creemers, J.H.A., Pawlitzky, I., Grosios, K. *et al.*, Assessing the safety, tolerability and efficacy of PLGA-based immunomodulatory nanoparticles in patients with advanced NY-ESO-1-positive cancers: a first-in-human phase I open-label dose-escalation study protocol. *BMJ Open*, 11, 11, e050725, 2021.
86. Makadia, H.K., Siegel, S.J., Poly Lactic-co-Glycolic Acid (PLGA) as Biodegradable Controlled Drug Delivery Carrier. *Polymers*, 3, 3, 1377–1397, 2011.
87. Petrovsky, N. and Cooper, P.D., Carbohydrate-based immune adjuvants. *Expert Rev. Vaccines*, 10, 4, 523–537, 2011.
88. Apostólico, J.D.S., Lunardelli, V.A.S., Coirada, F.C., Boscardin, S.B., Rosa, D.S., Adjuvants: Classification, *Modus Operandi*, and Licensing. *J. Immunol. Res.*, 2016, 1–16, 2016.
89. Reed, S.G., Orr, M.T., Fox, C.B., Key roles of adjuvants in modern vaccines. *Nat. Med.*, 19, 12, 1597–1608, 2013.
90. Mohammed, I., Nauman, A., Paul, P. *et al.*, The efficacy and effectiveness of the COVID-19 vaccines in reducing infection, severity, hospitalization, and mortality: a systematic review. *Hum. Vaccines Immunother.*, 18, 1, 2027160, 2022.
91. Sadoff, J., Le Gars, M., Shukarev, G. *et al.*, Interim Results of a Phase 1–2a Trial of Ad26.COV2.S Covid-19 Vaccine. *N. Engl. J. Med.*, 384, 19, 1824–1835, 2021.
92. Wang, Y., Li, J., Hu, Y., Liang, Q., Wei, M., Zhu, F., Ebola vaccines in clinical trial: The promising candidates. *Hum. Vaccines Immunother.*, 13, 1, 153–168, 2017.
93. Mohsen, M.O. and Bachmann, M.F., Virus-like particle vaccinology, from bench to bedside. *Cell. Mol. Immunol.*, 19, 9, 993–1011, 2022.
94. Zepeda-Cervantes, J., Ramírez-Jarquín, J.O., Vaca, L., Interaction Between Virus-Like Particles (VLPs) and Pattern Recognition Receptors (PRRs) From Dendritic Cells (DCs): Toward Better Engineering of VLPs. *Front. Immunol.*, 11, 1100, 2020.
95. Tariq, H., Batool, S., Asif, S., Ali, M., Abbasi, B.H., Virus-Like Particles: Revolutionary Platforms for Developing Vaccines Against Emerging Infectious Diseases. *Front. Microbiol.*, 12, 790121, 2022.
96. Gutjahr, A., Phelip, C., Coolen, A.-L. *et al.*, Biodegradable Polymeric Nanoparticles-Based Vaccine Adjuvants for Lymph Nodes Targeting. *Vaccines*, 4, 4, 34, 2016.
97. Bhadran, A., Shah, T., Babanyinah, G.K. *et al.*, Recent Advances in Polycaprolactones for Anticancer Drug Delivery. *Pharmaceutics*, 15, 7, 1977, 2023.
98. Chavez-Santoscoy, A.V., Roychoudhury, R., Pohl, N.L.B., Wannemuehler, M.J., Narasimhan, B., Ramer-Tait, A.E., Tailoring the immune response by

targeting C-type lectin receptors on alveolar macrophages using "pathogen-like" amphiphilic polyanhydride nanoparticles. *Biomaterials*, 33, 18, 4762–4772, 2012.
99. Strzelec, M., Detka, J., Mieszczak, P., Sobocińska, M.K., Majka, M., Immunomodulation—a general review of the current state-of-the-art and new therapeutic strategies for targeting the immune system. *Front. Immunol.*, 14, 1127704, 2023.
100. Sameer, A.S. and Nissar, S., Toll-Like Receptors (TLRs): Structure, Functions, Signaling, and Role of Their Polymorphisms in Colorectal Cancer Susceptibility. *BioMed Res. Int.*, 2021, 1–14, 2021.
101. Duan, T., Du, Y., Xing, C., Wang, H.Y., Wang, R.-F., Toll-Like Receptor Signaling and Its Role in Cell-Mediated Immunity. *Front. Immunol.*, 13, 812774, 2022.
102. Silva-Sanchez, A. and Randall, T.D., Anatomical Uniqueness of the Mucosal Immune System (GALT, NALT, iBALT) for the Induction and Regulation of Mucosal Immunity and Tolerance [Internet]. In: *Mucosal Vaccines*, pp. 21–54, Elsevier, 2020 [cited 2024 Apr 24]. Available from: https://linkinghub.elsevier.com/retrieve/pii/B978012811924200002X.
103. Neutra, M.R. and Kozlowski, P.A., Mucosal vaccines: the promise and the challenge. *Nat. Rev. Immunol.*, 6, 2, 148–158, 2006.
104. Bedford, J.G., Heinlein, M., Garnham, A.L. *et al.*, Unresponsiveness to inhaled antigen is governed by conventional dendritic cells and overridden during infection by monocytes. *Sci. Immunol.*, 5, 52, eabb5439, 2020.
105. Gouttefangeas, C. and Rammensee, H.-G., Personalized cancer vaccines: adjuvants are important, too. *Cancer Immunol. Immunother.*, 67, 12, 1911–1918, 2018.
106. Poland, G.A., Ovsyannikova, I.G., Jacobson, R.M., Personalized vaccines: the emerging field of vaccinomics. *Expert Opin. Biol. Ther.*, 8, 11, 1659–1667, 2008.
107. Kim, J.Y., Rosenberger, M.G., Rutledge, N.S., Esser-Kahn, A.P., Next-Generation Adjuvants: Applying Engineering Methods to Create and Evaluate Novel Immunological Responses. *Pharmaceutics*, 15, 6, 1687, 2023.
108. Da Silva, F.T., Di Pasquale, A., Yarzabal, J.P., Garçon, N., Safety assessment of adjuvanted vaccines: Methodological considerations. *Hum. Vaccines Immunother.*, 11, 7, 1814–1824, 2015.
109. Petrovsky, N., Comparative Safety of Vaccine Adjuvants: A Summary of Current Evidence and Future Needs. *Drug Saf.*, 38, 11, 1059–1074, 2015.

5

Adjuvants in Licensed Vaccines

Kaushika Patel[1*], Nandita Chawla[1], Yashvi Mehta[1] and Sachin Patel[2]

[1]Department of Pharmaceutical Technology, L. J. Institute of Pharmacy, LJ University, Ahmedabad, Gujarat, India
[2]F & D, OSD, Stallion Laboratories Pvt Ltd, India, Ahmedabad, Gujarat, India

Abstract
In the present circumstances, immunization is crucial as it ensures our safety and shields us from harmful pathogens. Although there have been significant achievements in the area of vaccinology, there is still room for progress in the development of safe and optimal vaccinations to combat life-threatening illnesses. Additionally, there is a need to enhance the effectiveness of current vaccines by addressing issues related to partial or variable protection. Therefore, it is quite likely that the potential of a vaccine may be enhanced by using innovative adjuvants, which can significantly enhance the immune response. The use of new adjuvants in vaccine formulation might potentially aid in combating diseases with significant antigenic diversity. Nevertheless, the approval of human-compatible adjuvants has been limited owing to safety and toxicity constraints. This study primarily highlights the need for novel and enhanced vaccinations, as well as the definition and requirement for adjuvants. This work explores several adjuvants used in vaccine development, including their mechanisms and targets. It puts a particular emphasis on the adjuvants used in COVID-19 and cancer vaccines. The adjuvants used in licensed vaccines are also summarized. The review has also included several advanced vaccination techniques and the toxicities associated with adjuvants.

Keywords: Vaccine, adjuvant, safety, toxicity, COVID-19, cancer, licensed vaccine

**Corresponding author*: kaushikaspatel@gmail.com

Vivek P. Chavda and Vasso Apostolopoulos (eds.) Emerging Pathways of Vaccine Adjuvants: A Nonspecific Stimulant of the Immune System, (107–136) © 2025 Scrivener Publishing LLC

5.1 Introduction

An adjuvant is an agent included into a vaccine to enhance the potency and persistence of the immune response inside the body [1]. The phrase adjuvant originates from the Latin term "adjuvare," which translates to "to assist or enhance." The term was invented by Ramon in 1920 [2].

Currently, the majority of inactivated vaccines are formulated using adjuvants, but the progress in this technology has been somewhat sluggish. Currently, only a limited number of adjuvants have been authorized for use in humans. In 1920, aluminum salts, sometimes known as alum, were the first adjuvant included in the formulation of tetanus and diphtheria toxoids. Subsequently, we had to wait until 1990 for the approval of the first vaccination that included a "novel" adjuvant called oil-in-water MF59, which was used in an influenza vaccine. In addition, a series of vaccines containing adjuvants were approved in the new millennium. These vaccines protect against avian influenza virus (with AS03), hepatitis B virus (HBV), human papillomavirus (HPV) (with AS04), herpes zoster virus (with AS01), and HBV (with CpG) [3–6].

There are various reasons for the comparatively unhurried development of novel adjuvanted vaccinations. The need for adjuvants has increased significantly due to the need for safer medication delivery, which involves the use of purer components such as purified recombinant antigens [7].

In contrast, vaccinations with lower immunogenicity, as opposed to live-attenuated or inactivated vaccines, have inherent adjuvant properties and are mediated by various immunostimulatory components. The composition of additional worth to substantiate the use of adjuvant above plain antigen is necessary for the showing of the enhanced value of adjuvant, hence lengthening the process of developing a vaccine [8, 9].

One of the most effective modes for preventing infectious illnesses is vaccination as said by World Health Organization (WHO). Conventional vaccination strategies, such as inactivated or live-attenuated viruses despite being extremely efficient and immunogenic, exhibit safety issues. Although subunit vaccines are safer yet are often less immunogenic/effective, they are supposed to be delivered along with an adjuvant to boost adaptive immune response to vaccine antigens [10, 11].

Adjuvants have become a crucial part of vaccinations because they may enhance the strength, duration, and effectiveness of the immune response to certain antigens, hence increasing the level and duration of protection afforded by the vaccines [12].

5.2 Adjuvants Included in Vaccines

Currently, there are many substances that have adjuvant characteristics and seem to work via several modes of action. Adjuvants may be categorized based on many factors to facilitate logical comparison. Adjuvants may be classified according to their physicochemical properties, origin, and mode of action [13].

Adjuvants may be categorized as delivery systems and immunological potentiators based on their mode of action. The third category, mucosal adjuvants, is such that it can fit into the previous two categories. Delivery mechanisms serve as carriers for the antigens that they carry. In addition, they stimulate the local pro-inflammatory responses that attract innate immune cells to the site of injection. Consequently, there has been a suggestion that this specific group of adjuvants may activate the body's natural defence mechanisms [14].

The immune potentiator targets innate immunity signalling pathways via Toll-like receptor (TLR), RIG-1 (retinoic acid-inducible gene-1)–like receptor (RLR), and NOD (nucleotide oligomerization domain)–like receptor (NLR). Typically, when PRRs are activated by their agonists, it leads to the activation and maturation of APCs, as well as the generation of cytokines and chemokines, which eventually result in adaptive immunological responses [12].

The summary of adjuvants along with their uses is enumerated in Table 5.1.

Table 5.1 Summary table of adjuvants and their utilization.

Category	Types	Adjuvants	Vaccines registered
Delivery system	Mineral salts	Aluminum salts (alum) and calcium phosphates	Hepatitis A (HAV), diphtheria/pertussis/tetanus (DPT), and human papillomavirus (HPV)
	Emulsion adjuvants	MF59 AS03	Influenza and in clinical trial for HPV and malaria
	Microparticles	Virus-like particles and virosomes Poly lactic acid (PLA)	Malaria, HPV, and HBV
Immune potentiators		CpG oligodeoxynucleotides (ODN)	HIV, hepatitis B, breast cancer, and hookworm infection
Combination adjuvants	AS04	HBV, HPV, HVA, DT, and DTaP	-
Mucosal adjuvants		Cholera toxin and chitosan	-

5.3 Cellular and Molecular Targets for Adjuvant

While adjuvants are used in several vaccines, their exact mechanism of action remains incompletely elucidated. Currently, their mechanism of action is gradually coming to light due to the studies from the last ten years. In this appraisal, we will be summarizing the latest advancements in our comprehension of the mechanism of action of adjuvants [15].

There are many possible methods by which adjuvants work, a few of them includes [15] the following:

- The development of depots or the controlled release of antigens at the site of injection
- Stimulation or enhancement of cytokines and chemokines
- Immune cell recruitment
- Improvement of the process by which antigens are taken in and presented to antigen-presenting cells (APCs)
- Facilitating the activation and maturation of APCs and enhancing the transportation of antigens to the lymph nodes that are responsible for emptying
- Activation of inflammasomes

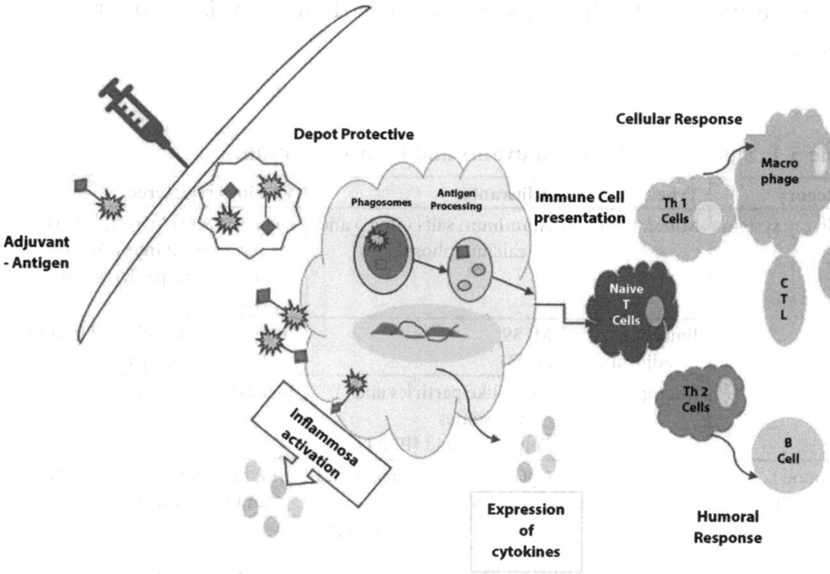

Figure 5.1 Molecular targets of adjuvants [12].

The Figure 5.1 shows the different molecular targets of the adjuvants used in vaccines. Adjuvants trigger the innate immune responses to induce a local immunological response at the injection site. Adjuvants have the capacity to alter the nature and extent of the adaptive immune response, based on the specific innate reactions they induce [16].

There are presently six adjuvants that are licensed and can be used in human vaccinations: alum, AS01, AS03, AS04, MF59, CpG, and ODN. This was achieved by the examination of various adjuvants' structure and the discovery of many pattern recognition receptors (PRRs) and co-stimulatory ligand receptors [17].

These findings have enhanced our comprehension of the molecular process behind the impact of adjuvants. To create vaccines that stimulate targeted immune responses and assess the safety of adjuvants throughout the several stages of development and regulation, it is crucial to understand how adjuvants work. In addition, we will examine the hypothetical processes by which vaccination adjuvant's function, with particular attention to approved adjuvants [18].

5.3.1 Depot Formation at the Injection Site

The most well-known and perhaps oldest mechanism of action for adjuvants is the creation of a depot at the injection site, wherein antigen trapping and slow release at the injection site constantly drive the immune system to produce high antibody titres [19].

The depot formation was primarily proposed from the adjuvant activity of alum. The straightforward adsorption of antigens onto the alum was believed to be caused by a robust electrostatic contact between the antigen and the alum, which improved antigen uptake and presentation by APCs. Additional adjuvants, including as biodegradable micro- and nanoparticles and water-in-oil emulsions, have been shown to function via the depot effect, resulting in prolonged and consistent production of high levels of antibodies [20].

AS04, an adjuvant composed of alum and monophosphoryl lipid A (MPL), has shown the ability to induce optimum immune responses only when it is present in the same location as an antigen. The availability of alum was found to be crucial for the stability of MPL and antigen in the vaccine and also to obtain a depot effect [16].

Depot effect used to be thought of as a standard mode of action for many adjuvants until recently. A depot effect was often assumed to be the only mechanism by which alum exerted its adjuvant function, but more recent research suggests that this is not the case. Also, MF59 does not require depot formation. Therefore, there is no definitive evidence that the depot effect has a substantial impact on adjuvant activity [20].

5.3.2 Induction and Upregulation of Cytokines and Chemokines

Current research has focused on the stimulation of innate immune cells at the injection site in relation to adjuvant processes. Evidence has shown that particle adjuvants are capable of triggering a localized pro-inflammatory state that activates immune cells [14].

Genome-wide microarray research demonstrated that the injection location of alum, MF59, and CpG-ODN usually caused alterations in a collection of genes that encode cytokines, chemokines, innate immune receptors, interferon-induced genes, and genes encoding adhesion molecules. Adjuvants have been discovered to express chemokines, which, in turn, play a crucial role to target immune cells to specific tissues [21].

It is difficult to interpret the activation of immune responses in relationship to immune cells; hence, an in-depth study of role of adjuvants is desirable [22].

5.3.3 Antigen Presentation

Antigen-presenting cells (APCs) contain major histocompatibility complexes (MHCs) to expose antigens for the elicitation of adaptive immune response.

Major histocompatibility complexes (MHCs) on APCs efficiently display antigens, which is essential for the development of the adaptive immune response. Alum is one of the adjuvants, which is known to strengthen the presentation of antigen in MHC to target the APCs [23]. The mechanism of adjuvant in inducing rises antigen presentation to produce adaptive response is an area yet to be explored further [24].

5.3.4 Activation and Maturation of DCs

The generation of an adaptive immune response involves DC activation. Improved expression of MHC class II, activation marker CD86, and maturation marker CD83 marks in an enhanced capacity of APCs to activate and differentiate T cells. In the end, adjuvants aid DCs mature and produce additional MHC and co-stimulatory molecules, which are vital for efficient T-cell activation [25].

5.3.5 Activation of Inflammasomes

Innate immune cells produce a number of pathogen-recognition receptors (PRRs) in order to identify dangerous infections. RLRs, TLRs, C-type lectin–like receptors, and NLRs are some of the novel families of PRRs that have been identified in recent investigations. Recently, an estimate of the inflammasome's contribution to MF59's adjuvant action was produced. Two separate investigations shown that, whereas ASC was discovered to be critical for MF59, NLRP4 is not required for the adjuvant action of MF59 [26].

Hence, the main objective of vaccination is to produce defences against microorganisms that cause disease. Immunity for defence against various pathogens demands various immune responses, which can be produced by utilizing the right adjuvants in vaccines. Thus, an in-depth understanding of adjuvants' modes of action is essential for the logical creation of vaccines [16].

5.4 Endogenous Adjuvants in Live Vaccines

A significant and successful strategy in public health to stop the spread of infectious illnesses is vaccination. According to the WHO, immunization is projected to save 20 million fatalities in nations with low or middle incomes from 2001 to 2020. The 280 billion US dollars in societal and economic benefits that result from this are significant. The advantages of vaccination in terms of saving lives and enhancing quality of life without impairments are indisputable. It is crucial to recognize that infectious illnesses continue to be the main cause of childhood death on a worldwide scale. This phenomenon can, in part, be attributed to the shortcomings of conventional vaccination approaches in preventing illnesses brought on by complex organisms [27].

Other issues in the area of vaccine development include maximizing vaccine production when confronted with inadequate antigen sources, boosting immune responses in populations with less receptive immune systems and enhancing the efficacy of pure recombinant antigens as immunogens. To address these challenges, innovative strategies and emerging technologies are being explored. These include viral vector vaccines, DNA vaccines, alternative administration routes like intradermal and intra-nasal methods, as well as novel adjuvants [28].

Adjuvants, which enhance the immune-stimulating properties of vaccines, play a pivotal role. For over half a century, aluminum was the solitary adjuvant used regularly in human vaccines, until recent developments [29].

5.4.1 Alum

Aluminum salts stand as the oldest, most extensively utilized and established safe adjuvants, having been in continual human use for nearly a century. We are currently experiencing an "aluminum era" where the availability of this metal has surged over the last 125 years, resulting in an increased overall aluminum load in humans. This surge in aluminum exposure primarily arises from environmental sources—our diet, drinking water, and air—as well as medical procedures, including vaccinations [30, 31].

Vaccination has played a pivotal role in improving both human and animal health, effectively eliminating significant diseases like smallpox and rinderpest. Since the establishment of the worldwide Alliance for Vaccination and Immunization in 2000 and the WHO's Expanded Programme of Immunization in 1974, there has been a worldwide decrease in measles infections and almost total poliomyelitis eradication. These vaccinations frequently contain dead or weakened pathogens, as well as poisons that serve as natural adjuvants. The use of aluminum salts in the early 1930s revealed a heightened immune response in humans compared to employing toxins alone in situations when toxins lack their own natural adjuvants, such as tetanus and diphtheria toxins [32–34].

As a result, vaccinations lacking their own integrated adjuvants have been combined with aluminum-based adjuvants (ABAs), including vaccines against toxins, pathogen subunits, and genetically engineered antigens [35–38].

Aluminum salts are frequently harnessed as adjuvants in vaccines, propelling and amplifying the immune system's reaction. The use of ABAs has a well-established history, yet concerns related to potential excessive vaccination and the reporting of unfavorable and harmful effects have emerged. The bulk of the known negative consequences linked to ABAs are self-tolerance disturbances, including the onset of autoimmune responses and the emergence of allergies. This is not unexpected considering that adjuvants are meant to boost the immune system's response, possibly disturbing the balanced defense mechanism that typically recognises pathogens without inducing self-antigens. Due to a molecular resemblance between vaccination antigens and self-antigens or due to an unrestrained immune system activation brought on by the adjuvant itself, this breach in self-tolerance and the initiation of autoimmunity may happen. The Autoimmune (Auto-inflammatory) Syndrome persuaded by Adjuvants (ASIA) is a name used to define a group of disorders, such as siliconosis, Gulf War syndrome, macrophagic myofasciitis (MMF), and post-vaccination abnormalities, which are triggered by adjuvants and result in harmful consequences. Gulf War syndrome, MMF, and post-vaccination symptoms are three ASIA disorders that may be exaggerated by ABAs, while documented negative effects from ABA use are scarce. The body's capability to hold ABAs and their potential for neurotoxicity are further problems. Aluminum ions have been demonstrated to hasten the production of amyloid, a protein linked to the onset of Alzheimer's disease, in both animals and humans [39].

The distribution of aluminum ions at both local and systemic levels, as well as the potential transportation of these ions across cell membranes, which might result in their buildup, is frequently discussed in discussions on risk assessment connected with ABAs in vaccines, particularly those that include aggregation of aluminum salts. Significantly, prolonged exposure to aluminum may lead to the formation of deposits in sentinel cells such as macrophages and dendritic cells (DCs), which are crucial components of the innate immune system. These phagocytic cells possess the capacity to induce inflammation, a crucial response that triggers the activation of the adaptive immune system and ultimately leads to the production of antibodies and cytotoxic T cells. Our knowledge of aluminum ion buildup after vaccinations delivered with ABA, however, is limited. Aluminum citrate has been used as a model to research aluminum distribution arising from the administration of ABAs, and this has provided the majority of insights into the distribution of aluminum ions in the body [40].

It is common to ignore the issue of adjuvant particle intracellular accumulation within phagocytic cells. The remaining cellular components will be swallowed by scavenger cells once these particles have been phagocytosed and the host cell has died. Given what is known about the longevity of ABAs in the body, it is plausible to assume that ABAs can be held inside these intracellular compartments for lengthy periods of time, ranging from months to even years, after vaccination [40].

5.4.2 Aluminum-Based Adjuvants

Glenny first introduced aluminum salts as vaccine adjuvants in 1926, and vaccination formulations have used them for almost a century. Despite their widespread use, the mechanisms behind ABAs' capacity to stimulate the immune system remain unknown [40].

The initial aluminum-adjuvanted vaccines were developed by precipitating antigens with aluminum potassium sulfate through the addition of a base. This process led to the formation of an antigen-aluminum salt complex. Aluminum potassium sulfate, commonly referred to as "alum," is incorrectly used to refer to all types of adjuvants made of aluminum. This precipitation approach was not successful due to limitations of batch reproducibility and variability in amount of adsorbed and entrapped antigen. The problems were overcome by adsorbing antigens on premade hydrated gels of aluminum salts to develop aluminum-adsorbed vaccines [41]. Alum is wrongly used to refer all forms of adjuvants produced of aluminum, because alum refers to only aluminum potassium sulfate [42].

The solubility minima for aluminum oxyhydroxide and aluminum hydroxy phosphate are in the pH range of 5–7. Aluminum oxyhydroxide's solubility rises significantly below pH 5, but aluminum hydroxy phosphate's solubility increases significantly below pH 6. Consequently, aluminum salts exist as aggregates of micron-sized particles both within a vaccine formulation and at the delivery site. The injection itself causes tissue injury, drawing sentinel cells from the local immune system. Aluminum ions have a low concentration and a sluggish translocation through cell membranes due to their restricted solubility. As a result, there is little intracellular aluminum buildup in the cells at the administration site. On the other hand, sentinel cells that are phagocytic invade the area and endocytose adjuvant particles, creating an intracellular aluminum reservoir. The phagosomes that contain these phagocytosed aluminum adjuvants develop into phagolysosomes, which are designed to break down ingested materials. However, because aluminum adjuvants are not biodegradable, they continue to exist as particles or aggregates inside the phagolysosome, where they act as a steady supply of aluminum ions [42, 43].

Aluminum adjuvant clumps are shown to remain inside these cells after being phagocytosed by macrophages at the injection site, according to *in vivo* studies. The adjuvant aggregates are transported by macrophages to draining lymph nodes, where they stay for years. Studies conducted *in vitro* employing cell lines show that cell division, not exocytosis, is responsible for decreases in intracellular aluminum levels. Adjuvant aggregates that have been phagocytosed persist intracellularly in differentiated macrophages. It is difficult to determine how long-lasting ABAs are *in vitro* because cell lines are constantly proliferating and diluting the accumulated adjuvant. Although they do not divide continuously, primary cells have a short *in vitro* lifetime, which limits the investigation of adjuvant persistence. Aluminum adjuvants may be tracked using the fluorescent marker lumiogallion after phagocytosis [40, 44, 45].

5.4.3 MF59

In order to increase the efficiency of vaccinations, adjuvants increase their capacity to elicit immunological responses. The oil-in-water emulsion MF59® has been licenced for use as an adjuvant for seasonal and pandemic influenza vaccines in several countries. Human populations have demonstrated MF59 to be safe and well-tolerated. Incorporating MF59 as an adjuvant enables vaccination dosage reduction while enhancing immune defenses against both varied and distantly related influenza virus strains. There are several different ways in which MF59 works its magic [21].

When MF59 is administered, a quick and significant synthesis of chemokines and inflammatory cytokines occurs. These chemicals function as signaling signals, luring different kinds of immune cells to the injection site. Furthermore, MF59 stimulates the production of uric acid, immune cell activation, and the safe cell death of certain innate immune cells. These coordinated efforts help to create a setting that supports efficient immune responses [21].

It is significant because MF59's adjuvant effects go beyond its capacity to produce isotype-switched immunoglobulin G (IgG) antibodies, which are important immune protection mediators. It also has a significant impact on effector CD8 T-cell proliferation, a subset of immune cells essential for directing immunological responses. MF59 creates an immunological response based on innate and APCs, which is effective even in the absence of CD4 T-cell subset, which is typically thought to be essential for coordinating immune responses [22].

This ability of MF59 to induce robust immunological responses in absence of CD4 T cells suggests its efficacy in different populations that not only include healthy adults but also cover children, the elderly, and immune-compromised people. The complex mechanisms by which MF59 acts as an adjuvant must be further explored by checking possibility of using other vaccine antigens and diverse demographic groups in order to develop even more successful vaccination techniques [46, 47].

5.4.4 Combination of Immune Stimulants: Adjuvant System (AS)

The combination of commonly used adjuvants with other substances that activate the immune system is preferred to generate desirable immunological response. Complicated formulations consisting of three or more adjuvant components have been developed to boost cellular immune response [48]. Such combination adjuvant systems are discussed in upcoming subsections.

5.4.4.1 AS04

The Adjuvant System 04 (AS04) is a combination adjuvant that has two components: a lipopolysaccharide (LPS) derivative known as monophosphoryl lipid A (MPL) and aluminum salts. Lipopolysaccharide (LPS), a potent immunostimulant and specific activator of TLR4, is produced by Gram-negative bacteria. MPL, a derivative of Salmonella minnesota's LPS, is generated by gently hydrolyzing and then purifying LPS. MPL has immune-stimulating properties similar to LPS, but without the adverse effects associated with original LPS. The AS04 formulation involves the adsorption of MPL onto either aluminum hydroxide or aluminum phosphate, depending on the specific vaccine being utilized [49].

MPL has a crucial role in activating the innate immune system in AS04. Antigen-presenting cells (APCs) undergo maturation due to the direct activation of TLR4, which also triggers the production of cytokines that enhance the adaptive immune response by promoting the maturation of T helper cells, namely those belonging to the Th1 subtype. Consequently, the recognition of MPL by TLR4 enhances more robust cellular and humoral immune responses. In order for AS04 to have an impact, it must be administered at the exact injection site as the antigen, either simultaneously or within a 24-h timeframe. As04 induces transient activation of innate immune cells and stimulates the release of cytokines throughout this period. This is shown by an increase in monocytes and activated DCs, which migrate toward the lymph nodes that receive the injection. Aluminum salts seem to control and prolong the cytokine responses to MPL at the injection site. Our findings confirm the notion that the inclusion of MPL in aluminum salt enhances the effectiveness of the vaccine by facilitating heightened activation of APCs and subsequently inducing Th1 T-cell responses [48].

5.4.4.2 AS03

AS03 is a mixture of adjuvants that contains squalene and vitamin E in an oil-in-water emulsion with droplets that are 150–155 nm in size. Vaccines against pandemic influenza use this combination. Vitamin E, a lipid-soluble antioxidant with immune-enhancing characteristics, is naturally present in several parts of the human body, including the muscles, adipose tissues, adrenal and pituitary glands, and pancreas. The primary role of this substance is to safeguard the structural integrity of cellular membranes and mitigate tissue damage caused by oxidative processes. Vegetable oils and a variety of nuts contain large amounts of vitamin E, which can only be produced by plants. It is widely used as a nutritional supplement in meals and is included in many cosmetic products. It is noteworthy that the vitamin E used in vaccinations is manufactured [50].

In response to AS03, monocytes and macrophages both cause the localized release of a variety of cytokines and chemokines. Typically, macrophages initiate cytokine response first, but recently recruited monocytes set off a subsequent wave of chemokine production that helps attract more innate immune cells [51].

5.4.4.3 AS01

A composite adjuvant system called AS01 uses QS21, liposomes, and MPL (a TLR4 agonist) in combination to maximize its synergistic effects. A South American tree's bark was used to extract the saponin that became the triterpene glycoside known as QS21. When evaluated in animal models, this adjuvant system showed promise in improving immune responses and boosting antigen presentation to APCs, especially in initiating the generation of cytotoxic T lymphocytes. Notably, QS21 can act as a surfactant to let proteins pass through cell membranes more easily, which triggers intracellular immunological reactions. In suitable formulations, QS21 has demonstrated acceptable tolerability profiles for use in human candidate vaccines. ISCOM, which stands for immune-stimulating complex consisting of cholesterol and phospholipids, and liposomes are two examples of such formulations [48].

The potential vaccine AS01, which combines liposomes, MPL, and QS21, is being investigated in a number of clinical trials for diseases like human immunodeficiency virus (HIV), TB, and malaria. One renowned example is the proposed malaria vaccine, which is now finalizing phase III clinical trials and aims the pre-erythrocytic stage of the malaria parasite using the recombinant antigen RTS, S. Both humoral and cell-mediated immune responses are required to decrease the number of infectious parasites that may attack the liver and eliminate infected hepatocytes [48].

5.4.4.4 AS15

AS15 is a multi-layered adjuvant system that synergistically blends the mechanisms of four different adjuvants: liposomes, MPL (a TLR4 agonist), CpG (a TLR9 agonist), and QS21. A study on the possible application of this elaborate concoction of adjuvants in cancer immunotherapy, namely, in the range of antigen-specific cancer immunotherapeutics (ASCI), is now continuing. In order to selectively target cancer cells though protecting the veracity of healthy cells, ASCI practices antigens that are either overexpressed or moderately expressed by tumor cells [52].

Due to AS15's unusual ability to intensify strong T-cell replies and prominent antibody creation, it was chosen for ASCI. In cancer immunotherapy, the substantial objective of AS15 is to boost the immune response against the target antigen. By combining a solid immune response, AS15 pursues to evade the immunosuppressive mechanisms that tumors frequently use. Clinical studies have discovered that AS15 may produce the extreme immune response for ASCI prerogatives when linked to other adjuvant combinations [52].

Different types of tumors express melanoma antigen A3 (MAGE-A3), which is the specific target of ASCI therapy using AS15. Positive conclusions from phase II studies have permitted for the use of MAGE-A3/AS15 as a cancer-specific immunotherapeutic approach in phase III clinical trials for melanoma and non–small-cell lung cancer. By successfully invading tumor cells with the immune system, this front-line adjuvant composition has the potential to transform the way cancer is treated [52].

5.4.5 Cytosine Phosphoguanosine 1018 (CpG 1018)

5.4.5.1 Saponin-Based Adjuvants

Immunostimulatory complexes (ISCOMs) are complex cage-like nanoparticle structures made from saponins isolated from the bark of Quillaja saponaria, a plant native to South America. These complexes are created using antigen, phospholipid, and cholesterol components. It is interesting to note that the inclusion of vaccination antigens within the particles is not required; in fact, present applications frequently combine soluble antigens with ISCOM particles that are devoid of antigens, as ISCOMATRIX.

ISCOMs have been shown to improve immunity without using any recognized PRRs. As opposed to this, they have a remarkable capacity to increase antigen absorption and lengthen the period of retention of antigens by DCs in the lymph nodes of the body. Additionally, ISCOMs cause DCs to become activated, which leads to potent antibody and T-cell responses. In particular, ISCOMs have the singular ability to support coordinated reactions from both Th1 and Th2 cells without predominating one over the other.

One of ISCOMs' unique features is its ability to allow for significant antigen presentation via MHC class I, effectively eliciting CD8+ and CD4+ T-cell responses to a range of soluble protein antigens in both experimental animal and humans. In contrast to other methods of antigen administration, ISCOMs appear to destabilize endosome membranes, allowing co-delivered antigens to have more access to the cytoplasm.

An ISCOM-derived heterogeneous saponin fraction known as Quil A is often used in the field of veterinary vaccinations. Furthermore, human experiments with multiple vaccine candidates and a highly purified variant of a particular saponin species known as QS-21 are now being conducted. This novel adjuvant strategy highlights the immunostimulatory capacity of ISCOMs and opens up exciting new directions for the production of vaccines [47].

5.4.5.2 Liposomal Adjuvants

Due to their particle form, liposomes with phospholipids as their primary lipid component have some intrinsic adjuvant effect. Due to features including depot effects and improved absorption by certain APCs, antigen-carrying particles frequently have immunostimulatory qualities. Today's adjuvants work via a range of concurrent properties and actions. Depot effects, delivery to specific cells, such as phagocytes, interactions with intracellular receptors or inflammasomes, immunity through TLR or non-TLR processes, presentation of danger signals indicating the presence of pathogens, triggers the release of particular cytokines or mediators, and preservation of membrane-associated antigens' natural conformational patterns [53].

Liposomal adjuvant compositions frequently include mixtures of ingredients because many types of liposomes have relatively limited adjuvant capability when used alone. These formulations comprise proteins, various immune-stimulators, stabilizers, and other substances either inside or outside of the liposomes that act as the structural scaffolding [54–56].

Research is being conducted on the development of subsequent liposomal adjuvants, including liposomes containing monophosphoryl lipid A (commonly referred to as the "Army Liposome Formulation") and the cationic liposomal formulation CAF01. Catalyst liposomes have been suggested as carriers for DNA delivery, even if the potential applications of DNA vaccines are not discussed here [57–59].

5.4.6 Adjuvants for Coronavirus Vaccines

Coronaviruses (CoVs), which are RNA viruses distinguished by spike proteins, can lead to serious respiratory conditions in people. The epidemics of SARS and Middle East respiratory disease (MERS) in 2002 have highlighted the critical need for effective immunizations. The necessity for the mass dissemination of effective immunizations has been further underscored by the COVID-19 pandemic that the SARS-CoV-2 virus is now causing [60, 61].

DNA vaccines, mRNA vaccines, adenovirus vaccines, live-attenuated vaccinations, recombinant protein vaccines, inactivated vaccines, and viral vector vaccines are a few of the vaccine platforms under development. Even though some of these platforms are new, all of them need strong adjuvants to guide and improve immune responses. In preclinical and clinical studies, adjuvants such aluminum salts, emulsions, and TLR agonists have been used with vaccine formulations for different animal models [62–64].

Adjuvants such AS03, MF59, and CpG 1018 that have been used in licensed vaccines are being investigated as part of the development of the COVID-19 vaccine. These adjuvants can alter the intensity, length, and direction of immune responses to create diverse immunological profiles. Many potential vaccines are now in the clinical and preclinical stages, and adjuvants are essential in determining how immunogenic they will be [65–68].

Alum has been used in conjunction with spike protein or receptor-binding domain (RBD) to produce antibodies that neutralize SARS-CoV-2; however, it may not be as efficient in triggering CD4+ and CD8+ T-cell responses. Other adjuvants, such as emulsion adjuvants and TLR agonists, are preferable options because they may stimulate both immune responses, namely, humoral and cellular. The precise protective function of various immune responses is still unclear, though, as no phase III clinical trial findings for COVID-19 vaccinations have been released [69–71].

Careful assessment of vaccine formulations and adjuvant selection is essential given the uniqueness of SARS-CoV-2 and the uncertainty surrounding its epidemiology and pathophysiology. The effectiveness of the vaccination depends on a number of adjuvant selection criteria, including antibody magnitude, affinity, isotype, and durability. In order to prevent antibody-dependent enhancement (ADE), it is crucial to avoid poor antibody production and to choose high-affinity neutralizing antibodies [72–74].

The utilized antigens also influence the adjuvant of choice. For instance, the full-length S protein may cause ADE by mildly producing antibodies, but the N protein may cause severe lung disease and the generation of pro-inflammatory cytokines. The antigen isotype may also influence the sort of immunological response that is induced. In order to provide the optimal immune protection against COVID-19, using the right adjuvants and antigens is crucial [75].

5.4.7 Cancer Vaccine Adjuvants

A potential approach called cancer immunotherapy makes use of the body's immune system to locate and destroy cancer cells. The immune system can fight off infections and illnesses thanks to its capacity to recognize alien or aberrant cells. As a type of pharmacological intervention, vaccinations have had great effectiveness in both preventing and treating a wide range of illnesses, both infectious and non-infectious. Despite the fact that vaccinations have significantly decreased mortality and morbidity, there is always a need for better tactics, particularly when it comes to difficult-to-treat infections like the HIV, hepatitis C virus, and respiratory syncytial virus [76].

The development of cancer vaccines has long been the focus of research and discussion. The success of recent clinical trials has increased interest in the creation of cancer vaccines. However, for cancer vaccines to be more effective, formulation changes must be made. One method to improve the calibre of immune responses induced by these vaccinations is the use of effective adjuvants. Adjuvants are compounds that are not naturally occurring in the body and that have a low level of toxicity or short-term immunological responses. These elements may occur in many chemical states, including liposomes, emulsions, surfactants, mineral salts, and toxic compounds that the immune system can recognize as cytokines, peptides, and receptor agonists [16, 77, 78].

The current efficacy revealed by cancer vaccines in clinical trials has created eagerness over their potential to efficiently treat and maybe eradicate many types of tumors. Subunit vaccinations, which precisely target certain components of tumors, have outdated the usage of whole tumor cells in the conception of cancer vaccines. The bulk of licenced or clinically tested vaccinations in the arena of cancer vaccine development are composed of subunit vaccines. This classification emphasises the need of leading attention on specific tumor antigens and using varied approaches to arouse immune responses against these antigens [79–83].

Including specific antigens obtained from the tumor is a crucial aspect of subunit vaccines. The selection of these antigens is based on their capacity to elicit an immune response against the cancer cells. To enhance the

Table 5.2 Adjuvants in licensed vaccines.

Adjuvants	Mode of action	Example of licensed vaccines
Aluminum salts	Depot effect and antigen persistence	Hepatitis B and human papillomavirus (HPV)
MF59	Oil-in-water emulsion	Influenza
AS04	Lipopolysaccharide derivative	Human papillomavirus (HPV) and hepatitis B,
AS03	Oil-in-water emulsion	Influenza
AS01	Liposome MPL	Malaria and shingles
ISCOMs	Saponins and lipids in particles	Experimental vaccines
QS21	Saponin	Malaria and shingles
Liposomal adjuvants	Liposome based formulation	Experimental vaccines

effectiveness of these subunit vaccines, other components such as DCs, DNA, and vectors (such as viral vectors) are often used with antigens [84, 85].

The identification of cancer vaccines has enhanced our understanding of how to precisely guide the immune system to selectively attack and eliminate cancerous cells. These strategies have the potential to advance cancer treatment and maybe enhance patient outcomes when research is conducted [86]. The adjuvants utilized in authorized vaccinations are outlined in Table 5.2.

5.5 Vaccine Adjuvants in COVID-19 Vaccines

The feedback to the SARS-CoV-2 pandemic has been majorly dependent on vaccine-related developments, testing, and deployments. Within a short timeframe, many vaccination methods were developed and then used on a large scale when their safety and effectiveness were confirmed.

These vaccines were administered to a significant population to safeguard against COVID-19 infections and deaths [72].

Prior to the emergence of novel vaccine designs, a significant challenge in the development of vaccines was that the simplicity and purification of antigens resulted in a subsequent loss in their ability to stimulate the immune system [87].

Amidst the pandemic, two vaccines using mRNA technology emerged and gained prominence in the United States: Pfizer/BioNTech BNT162b2 and Moderna mRNA-1273. Additionally, a vaccination that utilizes a viral vector as its delivery system [Jans-sen/J&J Ad26.COV2. S] was in use as well. Subsequently, a recombinant protein-based vaccine, i.e., Novavax NVX-CoV2373, also received approval and was in use. Thus, these four vaccines served as the representatives of three major vaccination platforms specifically, viral vector, mRNA, and recombinant protein plus adjuvant, which were utilized for the prevention of COVID-19 and offered more than 90% initial efficacy in the phase 3 trial [88–91].

5.5.1 Reasons and the Advantages of Adjuvant Incorporation Into Vaccines Against COVID-19

Adjuvants have been used for years in a number of vaccines against several infectious diseases, indicating their utility as well as enhancing the efficacy of the vaccine [47].

Some of the primary reasons for the adjuvant incorporation into the vaccine against COVID-19 are listed below [92]:

A. To increase vaccination response in the broader community, especially in the susceptible populations who respond less well to vaccination because of illness or aging.
B. For the reduction in the amount of antigen required for each dose of vaccine.
C. They enable immunization with fewer vaccine doses.
D. They trigger immune responses beyond antibodies.
E. They elicit heterotypic responses against different antigens.

5.5.2 Current Adjuvanted COVID-19 Vaccines

A vaccine against COVID-19 has to get beyond the ongoing variability of SARS-CoV-2 in order to be effective. It was discovered that one of the main issues in creating a vaccine against RNA-type viruses was the antigenic drift potential. This is because the RNA molecule is more prone to instability than the DNA molecule, which causes a higher rate of mutation and its frequent occurrence. Also, the lack of various polymerases in some viruses is responsible as well for the great antigenic and genetic drift against the SARS-CoV-2 virus [93].

Therefore, adjuvants must be added to older vaccine designs to improve both immunogenicity and utility if we are to remain ahead of these viruses and have a larger vaccine arsenal [94].

The ability of adjuvants to elicit a more complicated immune response based on innate immunity and a mix of humoral and cellular responses is one of the most exciting elements of using them with COVID-19 vaccines. While some of the more advanced vaccination designs may activate CD4, CD8, and B lymphocytes as well as elicit a more comprehensive response by drawing cells to the injection site, adjuvants may be essential in eliciting comparable responses in vaccines that are more traditional.

Table 5.3 lists a number of advanced vaccine designs that have been approved or are undergoing evaluation. These vaccines are recombinant protein vaccines that have several types of adjuvants included in their formulation [95].

Table 5.3 Advanced vaccine designs under review.

Vaccine name	Vaccine category	Adjuvant type	Clinical phase status
Novavax COVID-19 Omicron	Protein subunit	Matix-M1	Phase 3
Nuvaxovid (NVX-CoV2373)	Protein subunit	MATRIX-M	Authorized
VidPrevtyn beta	Protein subunit	AS03	Authorized
PHH-1V	Protein subunit	MF59	Authorized
VLA2001	Whole inactivated virus	CpG 1018	Authorized

5.6 Adjuvant-Related Toxicities

For more than 60 years, adjuvants have been employed in vaccinations and experimental pharmacology to enhance the immune response. Adjuvants have a diverse spectrum of chemical properties, methods of action, and side effects. Although some of the side effects may be the result of inadvertent immune system stimulation, others may be the result of more widespread negative pharmacological reactions. Currently, in choosing adjuvants for human vaccinations, a balance has been made between the need for adjuvanticity and an acceptable low degree of adverse effects [96].

Ideally, a vaccine adjuvant should increase immunogenicity yet preserve tolerance and safety. However, adjuvant research has unfortunately trailed behind other vaccine fields like antigen identification as a result a few vaccinations have received approval for use in humans [97].

Most of the vaccinations' efficiency depends on the adjuvants incorporated. Adjuvant research has long been focused on improving the drug's ability to induce immunogenicity; as a result, the assessment of its toxicity pathways has received less attention. The antigenic structure, dosages, and physicochemical properties of the adjuvants have the greatest impact on the quality of the elicited immune response. Likewise, the toxicity reactions are also influenced by these factors [98].

There are two types of responses elicited by the adjuvanted vaccines, i.e., local and systemic reactions.

5.6.1 Adjuvant-Associated Local Toxicity

Local reactions are the most recurring adverse effects of parenteral vaccines. These reactions might include lymphadenopathy, granulomas, and discomfort and swelling at the injection site. In spite of such high-frequency responses, the mechanisms behind such reactions are not evident yet [99].

Some local side effects such as irritation could be a reflection of abnormal pH, osmolarity, or maybe cell toxicity [100]. Such responses are associated with long-term repercussions followed by tissue damage [101].

5.6.2 Adjuvant-Associated Systemic Toxicity

Some systemic responses involved as a result of innate immune system activation are fever, headache, nausea, myalgia, lethargy, malaise, and diarrhoea. Therefore, adjuvants that stimulate the innate immune receptors are majorly prone to systemic side effects [100].

At the serious end of the systemic toxicity, adjuvants can potentiate a chronic immune activation and inflammation or may elicit organ toxicity of its compounds as well. Also, adjuvants raise the possibility or may trigger an autoimmune disease [102].

Henceforth, it can be said that, even after a century, aluminum salts have dominated as adjuvants, reflecting their high effectiveness and tolerability. Thus, in comparison to all new adjuvants, aluminum adjuvants remain the gold standard [97].

5.7 Conclusion

Vaccine adjuvants are essential components of vaccines that possess potential to increase the efficacy, longevity, and potency of the immune response to certain antigens. Apart from individual adjuvant, a combination system consisting of adjuvant with additional substances has been developed till date, to improve the potency and efficacy of vaccine. Due to the emergence of several types of vaccines, there is a huge scope in the advancement of adjuvants to be used in vaccine technology.

References

1. Pulendran B, S., Arunachalam, P., O'Hagan, D.T., Emerging concepts in the science of vaccine adjuvants. *Nat. Rev. Drug Discov.*, 20, 454–75, 2021, https://doi.org/10.1038/s41573-021-00163-y.
2. Sayers, S., Ulysse, G., Xiang, Z., He, Y., Vaxjo: A web-based vaccine adjuvant database and its application for analysis of vaccine adjuvants and their uses in vaccine development. *J. Biomed. Biotechnol.*, 2012, 831486, 2012, https://doi.org/10.1155/2012/831486.
3. Garçon, N. and Di Pasquale, A., From discovery to licensure, the Adjuvant System story. *Hum. Vaccin. Immunother.*, 13, 19–33, 2017, https://doi.org/10.1080/21645515.2016.1225635.
4. Guy, B., The perfect mix: Recent progress in adjuvant research. *Nat. Rev. Microbiol.*, 5, 505–17, 2007, https://doi.org/10.1038/nrmicro1681.
5. Vogel, F.R., Immunologic Adjuvants for Modern Vaccine Formulations. *Ann. N. Y. Acad. Sci.*, 754, 153–160, 1995.
6. Sepkowitz, K.A., The Myth of the Medical Breakthrough: Smallpox, Vaccination, and Jenner Reconsidered. *Int. J. Infect. Dis.*, 3, 1, 54–60, 1998.
7. McKee, A.S. and Marrack, P., Old and new adjuvants. *Curr. Opin. Immunol.*, 47, 44–51, 2017, https://doi.org/10.1016/j.coi.2017.06.005.

8. Del Giudice, G., Rappuoli, R., Didierlaurent, A.M., Correlates of adjuvanticity: A review on adjuvants in licensed vaccines. *Semin. Immunol.*, 39, 14–21, 2018, https://doi.org/10.1016/j.smim.2018.05.001.
9. Hutchison, S., Benson, R.A., Gibson, V.B., Pollock, A.H., Garside, P., Brewer, J.M., Antigen depot is not required for alum adjuvanticity. *FASEB J.*, 26, 1272–9, 2012, https://doi.org/10.1096/fj.11-184556.
10. Cain, D.W., Sanders, S.E., Cunningham, M.M., Kelsoe, G., Disparate adjuvant properties among three formulations of "alum". *Vaccine*, 31, 653–60, 2013, https://doi.org/10.1016/j.vaccine.2012.11.044.
11. Gavin, A.L., Hoebe, K., Duong, B., Ota, T., Martin, C., Beutler, B. *et al.*, Adjutant-enhanced antibody responses in the absence of toll-like receptor signaling. *Science (1979)*, 314, 1936–8, 2006, https://doi.org/10.1126/science.1135299.
12. Apostólico, J.D.S., Lunardelli, V.A.S., Coirada, F.C., Boscardin, S.B., Rosa, D.S., Adjuvants: Classification, Modus Operandi, and Licensing. *J. Immunol. Res.*, 2016, 1459394, 2016, https://doi.org/10.1155/2016/1459394.
13. Singh, M. and O'hagan, D.T., Recent Advances in Vaccine Adjuvants. *Pharm. Res.*, 19, 715–28, 2002.
14. Goto, N. and Akama, K., Histopathological Studies of Reactions in Mice Injected with Aluminum-Adsorbed Tetanus Toxoid. *Microbiol. Immunol.*, 26, 1121–32, 1982.
15. Cox, J.C. and Coulter, A.R., Adjuvants-a classification and review of their modes of action. *Vaccine*, 15, 248–56, 1997.
16. Awate, S., Babiuk, L.A., Mutwiri, G., Mechanisms of action of adjuvants. *Front. Immunol.*, 4, 114, 2013, https://doi.org/10.3389/fimmu.2013.00114.
17. García, A. and De Sanctis, J.B., An overview of adjuvant formulations and delivery systems. *APMIS*, 122, 257–67, 2014, https://doi.org/10.1111/apm.12143.
18. Sarkar, I., Garg, R., van Drunen Littel-van den Hurk, S., Selection of adjuvants for vaccines targeting specific pathogens. *Expert Rev. Vaccines*, 18, 505–21, 2019, https://doi.org/10.1080/14760584.2019.1604231.
19. Slsklnd, G.W. and Benacerraf, B., Cell Selection by Antigen in the Immune Response. *Adv. Immunol.*, 10, 1–50, 1969.
20. Mannhalter, J.W., Neychev, H.0., GJ, Z., Ahmad, R., Eibl, M.M., Modulation of the human immune response by the non-toxic and non-pyrogenic adjuvant aluminium hydroxide: effect on antigen uptake and antigen presentation. *Clin. Exp. Immunol.*, 61, 1, 143, 1985.
21. Mosca, F., Tritto, E., Muzzi, A., Monaci, E., Bagnoli, F., Iavarone, C. *et al.*, Molecular and cellular signatures of human vaccine adjuvants. *Proc. Natl. Acad. Sci.*, 105, 30, 10501–6. 2008.
22. Seubert, A., Monaci, E., Pizza, M., O'Hagan, D.T., Wack, A., The Adjuvants Aluminum Hydroxide and MF59 Induce Monocyte and Granulocyte Chemoattractants and Enhance Monocyte Differentiation toward Dendritic

Cells. *J. Immunol.*, 180, 5402–12, 2008, https://doi.org/10.4049/jimmunol.180.8.5402.

23. Guéry, J.-C., Ilia, F., Adorini, L., Dendritic Cells but Not B Cells Present Antigenic Complexes to Class II-restricted T Cells after Administration of Protein in Adjuvant. *J. Exp. Med.*, 183, 3, 751–7, 1996.
24. Ghimire, T.R., Benson, R.A., Garside, P., Brewer, J.M., Alum increases antigen uptake, reduces antigen degradation and sustains antigen presentation by DCs *in vitro*. *Immunol. Lett.*, 147, 55–62, 2012, https://doi.org/10.1016/j.imlet.2012.06.002.
25. Coyle, A.J. and Gutierrez-Ramos, J.-C., The expanding B7 superfamily: Increasing complexity in costimulatory signals regulating T cell function. *Nat. Immunol.*, 2, 3, 203–9, 2001.
26. Matzinger, P., Tolerance, danger, and the extended family. *Annu. Rev. Immunol.*, 12, 1991–1045, 1994.
27. Ozawa, S., Clark, S., Portnoy, A., Grewal, S., Stack, M.L., Sinha, A. *et al.*, Estimated economic impact of vaccinations in 73 low- and middleincome countries 2001–2020. *Bull. World Health Organ.*, 95, 629–38, 2017, https://doi.org/10.2471/BLT.16.178475.
28. Naghavi, M., Wang, H., Lozano, R., Davis, A., Liang, X., Zhou, M. *et al.*, Global, regional, and national age-sex specific all-cause and cause-specific mortality for 240 causes of death, 1990-2013: A systematic analysis for the Global Burden of Disease Study 2013. *Lancet*, 385, 117–71, 2015, https://doi.org/10.1016/S0140-6736(14)61682-2.
29. Laupèze, B., Hervé, C., Di Pasquale, A., Tavares Da Silva, F., Adjuvant Systems for vaccines: 13 years of post-licensure experience in diverse populations have progressed the way adjuvanted vaccine safety is investigated and understood. *Vaccine*, 37, 5670–80, 2019, https://doi.org/10.1016/j.vaccine.2019.07.098.
30. Rinella, J.V., Workman, R.F., Hermodson, M.A., White, J.L., Hem, S.L., Elutability of Proteins from Aluminum-Containing Vaccine Adjuvants by Treatment with Surfactants. *J. Colloid Interface Sci.*, 197, 48–56, 1998.
31. Johnston, C.T., Wang, S.-L., Hem, S.L., Measuring the Surface Area of Aluminum Hydroxide Adjuvant. *J. Pharm. Sci.*, 91, 1702–6, 2002.
32. May, J.C., The Muminum content of biological products containing aluminum adjuvants: determination by atomic absorption spectrometry. *J. Biol. Stand.*, 12, 2, 175–83, 1984.
33. Walls, R.S., Eosinophil Response to Alum Adjuvants: Involvement of T Cells in Non-Antigen-Dependent Mechanisms. *Proc. Soc. Exp. Biol. Med.*, 156, 431–5, 1977, https://doi.org/10.3181/00379727-156-39951.
34. Seeber, S.J., White, J.L., Hem, S.L., Predicting the adsorption of proteins by aluminium-containing adjuvants. *Vaccine*, 9, 3, 201–3, 1991.
35. Lindblad, E.B., Aluminium adjuvants - In retrospect and prospect. *Vaccine*, 22, 3658–68, 2004, https://doi.org/10.1016/j.vaccine.2004.03.032.

36. Fujimaki, H., Ozawa, M., Imai, T., Kuboto, K., Watanabe, N., International Archives of Allergy and Applied Immunology, 75, 4, 351–356, 1984, https://doi.org/10.1159/000233646.
37. OM, J. and Koch, C., On the effect of Al (OH)3 as an immunological adjuvant. *APMIS*, 96, 257–64, 1988, https://doi.org/10.1111/j.1699-0463.1988.tb05299.x.
38. He, P., Zou, Y., Hu, Z., Advances in aluminum hydroxide-based adjuvant research and its mechanism. *Hum. Vaccin. Immunother.*, 11, 477–88, 2015, https://doi.org/10.1080/21645515.2014.1004026.
39. Shirodkar, S., Hutchinson, R.L., Perry, D.L., White, J.L., Hem, S.L., Aluminum compounds used as adjuvants in vaccines. *Pharm. Res.*, 7, 12, 1282–8, 1990, doi: 10.1023/a:1015994006859. PMID: 2095567.
40. Danielsson, R. and Eriksson, H., Aluminium adjuvants in vaccines – A way to modulate the immune response. *Semin. Cell Dev. Biol.*, 115, 3–9, 2021, https://doi.org/10.1016/j.semcdb.2020.12.008.
41. Lu, F., Boutselis, I., Borch, R.F., HogenEsch, H., Control of antigen-binding to aluminum adjuvants and the immune response with a novel phosphonate linker. *Vaccine*, 31, 4362–7, 2013, https://doi.org/10.1016/j.vaccine.2013.07.019.
42. Rinella, J.V., White, J.L., Hem, S.L., Effect of pH on the Elution of Model Antigens from Aluminum-Containing Adjuvants. *J. Colloid Interface Sci.*, 205, 1, 161–5, 1998.
43. Lee, S. and Nguyen, M.T., Recent Advances of Vaccine Adjuvants for Infectious Diseases. *Immune Netw.*, 15, 51–7, 2015.
44. Danielsson, R., Svensson, A., Falkman, P., Eriksson, H., Tracing Aluminium-based Adjuvants: Their Interactions with Immune Competent Cells and their Effect on Mitochondrial Activity. *Open Immunol. J.*, 08, 1–15, 2018, https://doi.org/10.2174/1874226201808010001.
45. Mile, I., Svensson, A., Darabi, A., Mold, M., Siesjö, P., Eriksson, H., Al adjuvants can be tracked in viable cells by lumogallion staining. *J. Immunol. Methods*, 422, 87–94, 2015, https://doi.org/10.1016/j.jim.2015.04.008.
46. Ott, G., Barchfeld, G.L., Chernoff, D., Radhakrishnan, R., Van Hoogevest, P., Van Nest, G., MF59 Design and Evaluation of a Safe and Potent Adjuvant for Human Vaccines. *Pharm. Biotechnol.*, 277–96, 1995.
47. Coffman, R.L., Sher, A., Seder, R.A., Vaccine adjuvants: Putting innate immunity to work. *Immunity*, 33, 492–503, 2010, https://doi.org/10.1016/j.immuni.2010.10.002.
48. Garçon, N., Leroux-Roels, G., Cheng, W.F., Vaccine adjuvants. *Perspect. Vaccinol.*, 1, 89–113, 2011, https://doi.org/10.1016/j.pervac.2011.05.004.
49. Didierlaurent, A.M., Morel, S., Lockman, L., Giannini, S.L., Bisteau, M., Carlsen, H. *et al.*, AS04, an Aluminum Salt- and TLR4 Agonist-Based Adjuvant System, Induces a Transient Localized Innate Immune Response Leading to Enhanced Adaptive Immunity. *J. Immunol.*, 183, 6186–97, 2009, https://doi.org/10.4049/jimmunol.0901474.

50. Morel, S., Didierlaurent, A., Bourguignon, P., Delhaye, S., Baras, B., Jacob, V. *et al.*, Adjuvant System AS03 containing α-tocopherol modulates innate immune response and leads to improved adaptive immunity. *Vaccine*, 29, 2461–73, 2011, https://doi.org/10.1016/j.vaccine.2011.01.011.
51. Leroux-Roels, I., Leroux-Roels, G., Clement, F., Vandepapelière, P., Vassilev, V., Ledent, E. *et al.*, A phase 1/2 clinical trial evaluating safety and immunogenicity of a varicella zoster glycoprotein e subunit vaccine candidate in young and older adults. *J. Infect. Dis.*, 206, 1280–90, 2012, https://doi.org/10.1093/infdis/jis497.
52. Brichard, V.G. and Lejeune, D., GSK's antigen-specific cancer immunotherapy programme: Pilot results leading to Phase III clinical development. *Vaccine*, 25, B61–71, 2007, https://doi.org/10.1016/j.vaccine.2007.06.038.
53. Alving, C.R., Rao, M., Steers, N.J., Matyas, G.R., Mayorov, A.V., Liposomes containing lipid A: An effective, safe, generic adjuvant system for synthetic vaccines. *Expert Rev. Vaccines*, 11, 733–44, 2012, https://doi.org/10.1586/erv.12.35.
54. Beck, Z., Matyas, G.R., Jalah, R., Rao, M., Polonis, V.R., Alving, C.R., Differential immune responses to HIV-1 envelope protein induced by liposomal adjuvant formulations containing monophosphoryl lipid A with or without QS21. *Vaccine*, 33, 5578–87, 2015, https://doi.org/10.1016/j.vaccine.2015.09.001.
55. Beck, Z., Matyas, G.R., Alving, C.R., Detection of liposomal cholesterol and monophosphoryl lipid A by QS-21 saponin and Limulus polyphemus amebocyte lysate. *Biochim. Biophys. Acta, Biomembr.*, 1848, 775–80, 2015, https://doi.org/10.1016/j.bbamem.2014.12.005.
56. Fries, L.F., Gordont, D.M., Richardst, R.L., Egant, J.E., Hollingdale, M.R., Gross, M. *et al.*, 20852; and Departments of NMolecular Genetics and IlProtein Biochemistry. *Proc. Natl. Acad. Sci.*, 89, 358–62, 1992.
57. Alvlng, C.R., Rlchards, R.L., Moss, J., Alvlng, L.I., Clements, J.D., Shiba, T. *et al.*, Effectiveness of liposomes as potential carriers of vaccines: applications to cholera toxin and human malaria sporozoite antigen. *Vaccine*, 4, 3, 166–72, 1986.
58. Heppner, D.G., Gordon, D.M., Gross, M., Wellde, B., Leitner, W., Krzych, U. *et al.*, Safety, Immunogenicity, and Efficacy of Plasmodium falciparum Repeatless Circumsporozoite Protein Vaccine Encapsulated in Liposomes. *J. Infect. Dis.*, 174, 361–6, 1996.
59. Alving, C.R., Beck, Z., Matyas, G.R., Rao, M., Liposomal adjuvants for human vaccines. *Expert Opin. Drug Deliv.*, 13, 807–16, 2016, https://doi.org/10.1517/17425247.2016.1151871.
60. Fehr, A.R. and Perlman, S., Coronaviruses: An overview of their replication and pathogenesis, in: *Coronaviruses: Methods Protocols*, pp. 1–23, Springer, New York, 2015, https://doi.org/10.1007/978-1-4939-2438-7_1.
61. Docea, A.O., Tsatsakis, A., Albulescu, D., Cristea, O., Zlatian, O., Vinceti, M. *et al.*, A new threat from an old enemy: Re-emergence of coronavirus

(Review). *Int. J. Mol. Med.*, 45, 1631–43, 2020, https://doi.org/10.3892/ijmm.2020.4555.
62. Liu, C., Zhou, Q., Li, Y., Garner, L.V., Watkins, S.P., Carter, L.J. *et al.*, Research and Development on Therapeutic Agents and Vaccines for COVID-19 and Related Human Coronavirus Diseases. *ACS Cent. Sci.*, 6, 315–31, 2020, https://doi.org/10.1021/acscentsci.0c00272.
63. van Doremalen, N., Lambe, T., Spencer, A., Belij-Rammerstorfer, S., Purushotham, J.N., Port, J.R. *et al.*, ChAdOx1 nCoV-19 vaccine prevents SARS-CoV-2 pneumonia in rhesus macaques. *Nature*, 586, 578–82, 2020, https://doi.org/10.1038/s41586-020-2608-y.
64. Lasaro, M.O. and Ertl, H.C.J., New insights on adenovirus as vaccine vectors. *Mol. Ther.*, 17, 1333–9, 2009. https://doi.org/10.1038/mt.2009.130.
65. Koirala, A., Joo, Y.J., Khatami, A., Chiu, C., Britton, P.N., Vaccines for COVID-19: The current state of play. *Paediatr. Respir. Rev.*, 35, 43–9, 2020, https://doi.org/10.1016/j.prrv.2020.06.010.
66. McPherson, C., Chubet, R., Holtz, K., Honda-Okubo, Y., Barnard, D., Cox, M. *et al.*, Development of a SARS coronavirus vaccine from recombinant spike protein plus delta inulin adjuvant, in: *Vaccine Design: Methods and Protocols: Volume 1: Vaccines for Human Diseases*, vol. 1403, pp. 269–84, Humana Press Inc, 2016, https://doi.org/10.1007/978-1-4939-3387-7_14.
67. Jiang, S., Lu, L., Du, L., Development of SARS vaccines and therapeutics is still needed. *Future Virol.*, 8, 1–2, 2013, https://doi.org/10.2217/fvl.12.126.
68. Coleman, C.M., Liu, Y.V., Mu, H., Taylor, J.K., Massare, M., Flyer, D.C. *et al.*, Purified coronavirus spike protein nanoparticles induce coronavirus neutralizing antibodies in mice. *Vaccine*, 32, 3169–74, 2014, https://doi.org/10.1016/j.vaccine.2014.04.016.
69. Sun, L., Sogo, Y., Wang, X., Ito, A., Biosafety of mesoporous silica nanoparticles: a combined experimental and literature study. *J. Mater. Sci. - Mater. Med.*, 32, 1–5, 2021, https://doi.org/10.1007/s10856-021-06582-y.
70. Sun, B., Ji, Z., Liao, Y.P., Wang, M., Wang, X., Dong, J. *et al.*, Engineering an effective immune adjuvant by designed control of shape and crystallinity of aluminum oxyhydroxide nanoparticles. *ACS Nano*, 7, 10834–49, 2013, https://doi.org/10.1021/nn404211j.
71. Ruwona, T.B., Xu, H., Li, X., Taylor, A.N., Shi, Y.C., Cui, Z., Toward understanding the mechanism underlying the strong adjuvant activity of aluminum salt nanoparticles. *Vaccine*, 34, 3059–67, 2016, https://doi.org/10.1016/j.vaccine.2016.04.081.
72. Jackson, L.A., Anderson, E.J., Rouphael, N.G., Roberts, P.C., Makhene, M., Coler, R.N. *et al.*, An mRNA Vaccine against SARS-CoV-2 — Preliminary Report. *N. Engl. J. Med.*, 383, 1920–31, 2020, https://doi.org/10.1056/nejmoa2022483.
73. Tseng, C.T., Sbrana, E., Iwata-Yoshikawa, N., PC, N., Garron, T., RL, A. *et al.*, Immunization with SARS coronavirus vaccines leads to pulmonary

immunopathology on challenge with the SARS virus. *PLoS One*, 7, e35421, 2012, https://doi.org/10.1371/journal.pone.0035421.
74. Takasuka, N., Fujii, H., Takahashi, Y., Kasai, M., Morikawa, S., Itamura, S. *et al.*, A subcutaneously injected UV-inactivated SARS coronavirus vaccine elicits systemic humoral immunity in mice. *Int. Immunol.*, 16, 1423–30, 2004, https://doi.org/10.1093/intimm/dxh143.
75. Liang, Z., Zhu, H., Wang, X., Jing, B., Li, Z., Xia, X. *et al.*, Adjuvants for Coronavirus Vaccines. *Front. Immunol.*, 11, 589833, 2020, https://doi.org/10.3389/fimmu.2020.589833.
76. Leroux-Roels, G., Unmet needs in modern vaccinology. Adjuvants to improve the immune response. *Vaccine*, 28, C25–36, 2010, https://doi.org/10.1016/j.vaccine.2010.07.021.
77. Pashine, A., Valiante, N.M., Ulmer, J.B., Targeting the innate immune response with improved vaccine adjuvants. *Nat. Med.*, 11, s63, 2005, https://doi.org/10.1038/nm1210.
78. Davis, I.D., Jefford, M., Parente, P., Cebon, J., Rational approaches to human cancer immunotherapy. *J. Leukoc. Biol.*, 73, 3–29, 2003, https://doi.org/10.1189/jlb.0502261.
79. Butterfield, L.H., Cancer vaccines. *BMJ (Online)*, 350, 2015, https://doi.org/10.1136/bmj.h988.
80. Harris, J.E., Ryan, L., Hoover, H.C., *et al.*, Adjuvant Active Specific Immunotherapy for Stage II and III Colon Cancer With an Autologous Tumor Cell Vaccine: Eastern Cooperative Oncology Group Study E5283. *J. Clin. Oncol.*, 18, 1, 2000, https://doi.org/10.1200/JCO.2000.18.1.148
81. Sioud, M., An Overview of the Immune System and Technical Advances in Tumor Antigen Discovery and Validation. *Target Discovery and Validation Reviews and Protocols: Volume 1, Emerging Strategies for Targets and Biomarker Discovery*, pp. 277–318, 2007.
82. Stevenson, F.K., Ottensmeier, C.H., Johnson, P., Zhu, D., Buchan, S.L., Mccann, K.J. *et al.*, DNA vaccines to attack cancer. *Proc. Natl. Acad. Sci.*, 101(suppl_2), 14646–52, 2004.
83. Garnett, C.T., Greiner, J.W., Tsang, K.-Y., Kudo-Saito, C., Grosenbach, D.W., Chakraborty, M. *et al.*, TRICOM Vector Based Cancer Vaccines. *Curr. Pharm. Des.*, 12, 351–61, 2006.
84. Dubensky, T.W. and Reed, S.G., Adjuvants for cancer vaccines. *Semin. Immunol.*, 22, 155–61, 2010, https://doi.org/10.1016/j.smim.2010.04.007.
85. Ragupathi, G., Gathuru, J., Livingston, P., Antibody inducing polyvalent cancer vaccines. *Tumor Immunology and Cancer Vaccines*, vol. 7, pp. 157–80, 2005.
86. Banday, A.H., Jeelani, S., Hruby, V.J., Cancer vaccine adjuvants - Recent clinical progress and future perspectives. *Immunopharmacol. Immunotoxicol.*, 37, 1–11, 2015, https://doi.org/10.3109/08923973.2014.971963.

87. Castrodeza-Sanz, J., Sanz-Muñoz, I., Eiros, J.M., Adjuvants for COVID-19 Vaccines. *Vaccines (Basel)*, 11, 902, 2023, https://doi.org/10.3390/vaccines11050902.
88. Vogel, A.B., Kanevsky, I., Che, Y., Swanson, K.A., Muik, A., Vormehr, M. *et al.*, BNT162b vaccines protect rhesus macaques from SARS-CoV-2. *Nature*, 592, 283–9, 2021, https://doi.org/10.1038/s41586-021-03275-y.
89. Novavax COVID19 Vaccine, Adjuvanted (2023-2024 Formula) HCP Fact Sheet. *Vaccines*, 12, 1, 74, 2024.
90. Pollard, A.J. and E.M., Bijker A guide to vaccinology: from basic principles to new developments. *Nat. Rev. Immunol.*, 21, 83–100, 2021, https://doi.org/10.1038/s41577-020-00479-7.
91. Pilz, S., Theiler-Schwetz, V., Trummer, C., Krause, R., Ioannidis, J.P.A., SARS-CoV-2 reinfections: Overview of efficacy and duration of natural and hybrid immunity. *Environ. Res.*, 209, 112911, 2022, https://doi.org/10.1016/j.envres.2022.112911.
92. Banzhoff, A., Gasparini, R., Laghi-Pasini, F., Staniscia, T., Durando, P., Montomoli, E. *et al.*, MF59®-adjuvanted H5N1 vaccine induces immunologic memory and heterotypic antibody responses in non-elderly and elderly adults. *PLoS One*, 4, e4384, 2009, https://doi.org/10.1371/journal.pone.0004384.
93. Peck, K.M. and Lauring, A.S., Complexities of Viral Mutation Rates 2018. *J. Virol.*, 94, 22, 10–128, 2020, https://doi.org/10.1128/JVI.
94. Manzanares-Meza, L.D. and Medina-Contreras, O., SARS-CoV-2 and influenza: A comparative overview and treatment implications. *Bol. Med. Hosp. Infant Mex.*, 77, 262–73, 2020, https://doi.org/10.24875/BMHIM.20000183.
95. Guo, M., Li, J., Wang, Y., Chen, G., Chen, R., Wang, L., The association between influenza vaccination and the perception of COVID-19 as well as COVID-19 vaccination behavior among community residents in Anhui province, China. *Hum. Vaccin. Immunother.*, 19, 2275464, 2023, https://doi.org/10.1080/21645515.2023.2275464.
96. Gupta, R.K., Relyveldt, E.H., Lindblad, E.B., Bizzini, B., Ben-Efraim, S.I., Kanta Gupta, C., Review Adjuvants a balance between toxicity and adjuvanticity. *Vaccine*, 11, 3, 293–306, 1993.
97. Petrovsky, N., Comparative Safety of Vaccine Adjuvants: A Summary of Current Evidence and Future Needs. *Drug Saf.*, 38, 1059–74, 2015, https://doi.org/10.1007/s40264-015-0350-4.
98. Batista-Duharte, A., Portuondo, D., Carlos, I.Z., Pérez, O., An approach to local immunotoxicity induced by adjuvanted vaccines. *Int. Immunopharmacol.*, 17, 526–36, 2013, https://doi.org/10.1016/j.intimp.2013.07.025.
99. Batista-Duharte, A., Lindblad, E.B., Oviedo-Orta, E., Progress in understanding adjuvant immunotoxicity mechanisms. *Toxicol. Lett.*, 203, 97–105, 2011, https://doi.org/10.1016/j.toxlet.2011.03.001.

100. Petrovsky, N., Freeing vaccine adjuvants from dangerous immunological dogma. *Expert Rev. Vaccines*, 7, 7–10, 2008, https://doi.org/10.1586/14760584.7.1.7.
101. Julio César, A., Vaccine adjuvants: Current state and future trends. *Immunol. Cell Biol.*, 82, 488–96, 2004, https://doi.org/10.1111/j.1440-1711.2004.01272.x.
102. Powell, B.S., Andrianov, A.K., Fusco, P.C., Polyionic vaccine adjuvants: another look at aluminum salts and polyelectrolytes. *Clin. Exp. Vaccine Res.*, 4, 23, 2015, https://doi.org/10.7774/cevr.2015.4.1.23.

6

Nanomaterial-Based Vaccine Adjuvants

Tanvi[1], Philips Kumar[1], Rajat Goyal[1], Kashish Wilson[1], Hitesh Chopra[2*] and Rajeev K. Singla[3,4†]

[1]MM College of Pharmacy, Maharishi Markandeshwar (Deemed to be University), Mullana-Ambala, Haryana, India
[2]Department of Biosciences, Saveetha School of Engineering, Saveetha Institute of Medical and Technical Sciences, Chennai, Tamil Nadu, India
[3]Joint Laboratory of Artificial Intelligence for Critical Care Medicine, Department of Critical Care Medicine and Institutes for Systems Genetics, Frontiers Science Center for Disease-Related Molecular Network, West China Hospital, Sichuan University, Chengdu, China
[4]School of Pharmaceutical Sciences, Lovely Professional University, Phagwara, Punjab, India

Abstract

This book chapter explores nanomaterials' transformative role in enhancing vaccine efficacy by designing and utilizing novel vaccine adjuvants. Nanomaterial-based vaccine adjuvants have emerged as a promising avenue for enhancing the immunogenicity of vaccines. Their unique physicochemical properties, such as size, shape, surface charge, and composition, enable them to interact with the immune system and modulate immune responses effectively. This chapter provides a comprehensive overview of the recent advances in developing and applying nanomaterial-based vaccine adjuvants. The chapter begins by discussing the fundamentals of vaccine adjuvants, their mechanisms of action, and the challenges associated with their development. Subsequently, the chapter delves into the various types of nanomaterials explored as vaccine adjuvants, including inorganic nanoparticles, organic nanoparticles, and hybrid nanomaterials. The chapter also discusses the different strategies for designing and optimizing nanomaterial-based vaccine adjuvants. The chapter then reviews the preclinical and clinical studies evaluating nanomaterial-based vaccine adjuvants' efficacy in various infectious

**Corresponding author*: chopraontheride@gmail.com
†*Corresponding author*: rajeevsingla26@gmail.com; rajeevkumar@scu.edu.cn

Vivek P. Chavda and Vasso Apostolopoulos (eds.) Emerging Pathways of Vaccine Adjuvants: A Nonspecific Stimulant of the Immune System, (137–170) © 2025 Scrivener Publishing LLC

diseases, such as cancer and autoimmune diseases. The chapter also highlights the potential of nanomaterial-based vaccine adjuvants for developing next-generation vaccines.

Keywords: Nanomaterials, vaccine adjuvants, immunogenicity, immune modulation, vaccine development

6.1 Introduction

In the dynamic field of vaccine development, finding safer and more effective adjuvants is crucial. Traditional vaccines often need adjuvants for enhanced efficiency, driving research and innovation. A recent focus on nanomaterial-based adjuvants shows promise, signifying a shift in vaccine design. Nanomaterials offer advantages like better antigen transport, increased immunogenicity, and precise immune response control. This revolutionary approach can effectively transform vaccine development and address infectious diseases and health challenges [1].

A particularly fascinating and promising development has just come into view: vaccination adjuvants based on nanomaterials. This innovative method represents a significant change in how we approach the creation and administration of vaccines. Due to their distinctive physicochemical characteristics and adaptability, nanomaterials provide a wide range of benefits that have attracted the interest of researchers and scientists worldwide [2]. The application of nanoparticles as vaccine adjuvants can transform the industry entirely. These nanometer-sized structures open a whole new world of opportunities for improving vaccines. Their capacity to enhance immunogenicity, increase antigen delivery, and accurately control immune responses opens the door to a future in which vaccines are more efficient and precisely adapted to address diseases and health issues [3]. These developments have the potential to fundamentally alter the way vaccines are developed, giving us the ability to combat infectious diseases and other health risks with unparalleled accuracy and efficacy. Nanomaterial-based vaccination adjuvants shine as a beacon of hope, illuminating the route toward a safer and healthier society in the ever-evolving healthcare scene [4].

6.1.1 Unveiling the Essence: Navigating Vaccine Definition and Conceptualizing Vaccines

A vaccine is a biological or pharmaceutical substance that boosts the body's immune system against specific diseases. Typically containing weakened, inactivated, or fragments of disease-causing microorganisms or synthetic molecules, vaccines prompt an immune response without causing the actual disease. This response, involving antibody formation and immune cell activation, provides immunity and reduces the risk and severity of future infections. Vaccines are crucial for preventing and controlling infectious diseases, significantly improving public health by reducing morbidity, mortality, and the spread of pathogens [5].

Nanomaterial-based vaccine adjuvants are revolutionizing vaccination by altering the fundamental nature of immunity. This chapter explores the transformative impact of nanomaterials on our approach to vaccines.

6.1.1.1 Nanomaterials as Immune Modulators

The ability of nanomaterials to operate as immune modulators is central to this approach. These materials, which are frequently nanoscale in size, have features that allow them to interact with the immune system in unique ways. Their size, shape, and surface qualities can all be customized to impact immune response amounts and direction. They can increase immune cell uptake of antigens, prolong antigen exposure, and steer immunological responses toward desired outcomes. This level of precision can revolutionize vaccine design, enabling us to develop vaccines that target a broader spectrum of infections, give longer-lasting protection, and eliminate the need for frequent booster shots.

6.1.1.2 Enhancing Efficacy and Safety

Adjuvants based on nanomaterials boost immune response and increase vaccine safety. They minimize the likelihood of adverse responses while increasing protection by lowering the antigen required in a vaccination. It is instrumental in the creation of vaccines for vulnerable groups such as babies, the elderly, and those with impaired immune systems. Furthermore, nanomaterials' controlled and predictable immune regulation promises to reduce vaccine-associated adverse effects, thus improving immunization's overall safety and acceptability.

6.1.1.3 Personalized Vaccinology

Vaccinology can be treated more individually due to the adaptability of nanomaterial-based adjuvants. Vaccinations can be designed to elicit specific immune responses, allowing researchers to develop vaccines for personalized protection against various infectious diseases. Individualized vaccinology opens new avenues for precision therapy and rapid response to emerging infectious threats. It allows us to envision a future in which vaccinations may be developed swiftly and targeted to specific viruses, potentially changing our ability to combat pandemics [6].

6.1.2 Importance in Preventing Infectious Diseases

1. **Global Health Challenges:** Infectious illnesses continue to offer crucial global health issues, with new pathogens emerging and existing ones persisting, inflicting illness, death, and economic upheaval.
2. **Limitations of Traditional Vaccines:** Traditional vaccinations have helped reduce illness incidence, but they have limitations, such as the requirement for regular booster shots and difficulties in successfully targeting specific pathogens.
3. **Precision and Potency:** Precision and potency are advantages of nanomaterial-based vaccine adjuvants. They can be precisely regulated to ensure vaccinations elicit the correct type and level of protection, even against complicated infections.
4. **Rapid Response to Emerging Threats:** Adjuvants based on nanomaterials enable quick vaccine production, which is critical in reacting to newly developing viruses such as SARS-CoV-2, responsible for the COVID-19 pandemic.
5. **Expanding the Vaccine Arsenal:** These adjuvants have the potential to broaden the vaccine arsenal by providing a versatile platform for producing vaccines against a wide range of infectious diseases, including some that have proven challenging to target using traditional approaches.
6. **Shaping the Future of Preventive Medicine:** Finally, nanomaterial-based vaccine adjuvants have the potential to alter preventive medicine. They provide precision, potency, and flexibility, hoping for a healthier and more resilient global health future as we face the challenges of infectious pathogens [7].

6.2 Vaccine Adjuvants and Their Role in Enhancing Immune Responses

Vaccine adjuvants are essential to modern vaccinology, enhancing and guiding the body's immunological reactions to immunizations. They are chemicals added to vaccinations to increase effectiveness by inducing a more substantial and longer-lasting immune response. These discoveries offer a substantial advancement in our ability to fine-tune the immune system's reactions for improved vaccine outcomes in the setting of nanomaterial-based vaccine adjuvants.

Here, we explain the concept of vaccination adjuvants and their critical function in increasing immune response, mainly when nanomaterials are included:

1. **Enhancing Immune Responses:** The fundamental function of vaccine adjuvants, whether traditional or nanomaterial-based, is to enhance the body's immunological response to the antigens of the vaccination. They accomplish this by simulating the presence of an actual infection and so more effectively activating immune cells.
2. **Complementing Antigens:** Adjuvants supplement the vaccine's antigens, often weakened or inactivated microorganisms. While antigens trigger the immune response, adjuvants strengthen and direct it, making it more powerful.
3. **Immunomodulation:** Nanomaterial-based adjuvants excel in fine-tuning the immune system because of their features. Because they can be precisely tailored to modify the type and severity of the immune response, they are invaluable tools for tailoring vaccine outcomes.
4. **Prolonged Immunity:** Adjuvants, particularly nanomaterials, help to prolong immunity by improving the immune system's ability to "remember" antigens. This memory results in a more robust response against future encounters with the virus.
5. **Reducing Vaccine Doses:** Adjuvants based on nanomaterials are frequently used to generate vaccinations that require lower antigen dosages, lowering the likelihood of adverse responses and optimizing vaccine safety.
6. **Targeting Challenging Pathogens:** Nanomaterial-based adjuvants provide new promise for pathogens that have

historically been difficult to target effectively with standard vaccinations. They improve the precision and efficacy of vaccine reactions, even against complicated infections.

7. **Rapid Vaccine Development:** The versatility of nanomaterial-based adjuvants allows for the quick generation of vaccines, which is critical in reacting to emerging disease threats. This skill has been instrumental during pandemics [8].

6.2.1 Mechanism of Action

Vaccine adjuvants play a crucial role in enhancing immune responses, contributing to the overall efficacy of immunization. Understanding their mechanisms sheds light on the intricate dynamics of vaccine-induced immunity.

1. **Activation of Innate Immune Sensors:** Adjuvants such as aluminum salts (alum) initiate the immune response by activating innate immune sensors like the NLRP3 inflammasome [9].
2. **Promotion of Antigen Presentation:** Through their depot effect, adjuvants create a sustained release of antigens, promoting prolonged interaction with antigen-presenting cells (APCs) [10].
3. **Induction of Cytokine Release:** Toll-like receptor (TLR) agonists, like monophosphoryl lipid A (MPL), stimulate APCs to produce pro-inflammatory cytokines, enhancing the immune response [11].
4. **Stimulation of Memory T Cells:** Adjuvants like the oil-in-water emulsion MF59 enhance the generation of memory T cells, contributing to prolonged protection against pathogens [12].
5. **Modulation of Antibody Responses:** Adjuvants, including CpG oligodeoxynucleotides, modulate antibody isotype switching, resulting in a diversified and effective humoral immune response [13].

6.2.2 Types of Adjuvants

Different adjuvants are used in vaccine production to increase the potency of vaccines. Based on their makeup and modes of action, these adjuvants can be divided into several groups. Here are a few prevalent vaccination adjuvants:

1. **Aluminum Salts (Alum):** Vaccines have long included aluminum-based adjuvants such as aluminum hydroxide and aluminum phosphate. They function by establishing a depot at the site of injection, which releases the antigen gradually and triggers an immunological response. Alum adjuvants used in vaccinations for hepatitis A, hepatitis B, and the human papillomavirus (HPV) are well-known for their safety profile.
2. **Oil-in-Water Emulsions:** Oil droplets distributed in a water-based solution make up these adjuvants. At the injection site, they have a depot effect that prolongs the antigen's stay and strengthens the immunological response. Examples include AS03 and MF59, used in various influenza vaccinations (used in some pandemic influenza vaccines) [14, 15].
3. **Toll-like receptor (TLR) Agonists:** TLR agonists imitate the infections' molecular patterns, turning on immune cell receptors and eliciting an immunological response. Mono phosphoryl lipid A (MPLA), a component of the AS04 adjuvant used in HPV and hepatitis B vaccinations, is one type of TLR agonist used as an adjuvant.
4. **Saponins:** Saponins are organic substances that can improve an immunological response. QS-21, a prominent saponin-based adjuvant utilized in creating vaccines for diseases including malaria and shingles, is generated from the bark of the *Quillaja saponaria* tree.
5. **Virosomes:** Lipid vesicles called virosomes contain viral or bacterial components. They can improve the immunological response by making it easier for antigens to reach immune cells. Influenza vaccinations have utilized viral adjuvants.
6. **Protein Subunit Adjuvants:** When combined with antigens, some adjuvants comprise protein subunits such as bacterial or viral proteins that boost the immune response. Bacterial toxoids (for example, tetanus toxoid) and viral-like particles [e.g., HPV virus-like particles (VLPs)] are examples of these.
7. **Cationic Liposomes:** Cationic liposomes are positively charged lipid-based vesicles. They can interact with negatively charged antigens, increasing antigen absorption by APCs and strengthening the immune response.
8. **Nanoparticles:** Nanoparticles, which are often comprised of lipids or polymers, can be designed to transport antigens and adjuvant components. They provide precise antigen delivery

and immunological activation. Lipid nanoparticles, for example, are employed in several COVID-19 vaccinations.

9. **Mineral Salts:** Calcium phosphate, for example, can form complexes with antigens, enhancing their stability and immunogenicity. These are frequently found in veterinary vaccinations.
10. **Combination Adjuvants:** Vaccines may contain a combination of adjuvants in some situations to maximize the immune response. The AS04 adjuvant, for example, comprises aluminum hydroxide and MPLA [16].

6.2.3 Need for Novel Adjuvants to Improve Vaccine Effectiveness and Safety

The need for novel adjuvants in vaccine development is driven by several crucial factors that revolve around improving vaccine effectiveness and safety. While traditional adjuvants like aluminum salts have been essential in enhancing vaccine responses, there are ongoing challenges and emerging considerations that necessitate the exploration of new adjuvant technologies:

1) **Enhancing Immunogenicity:** Some pathogens, such as certain viruses or bacteria, may have weak antigens that do not elicit a robust immune response. Novel adjuvants are required to boost the immunogenicity of these antigens, ensuring that vaccines provide robust and long-lasting protection [17].
2) **Improving Efficacy in Vulnerable Populations:** Certain populations, such as the elderly and those with impaired immune systems, may have lower vaccine responses. New adjuvants can assist in overcoming these restrictions by boosting immune responses, allowing vulnerable people to benefit from vaccination [18].
3) **Reducing Antigen Dose:** Reduced antigen content in vaccination can help to reduce potential side effects and production costs. New adjuvants may allow lower antigen dosages while preserving or boosting immunization effectiveness [19].
4) **Expanding Vaccine Coverage:** Adjuvants can be customized to specific pathogens or diseases, allowing vaccines to be developed against difficult or previously unreachable

targets. This increase in vaccination coverage is critical for combating emerging infectious illnesses and existing unmet medical requirements [20].

5) **Creating Safer Vaccines:** Safety is of the utmost importance in vaccine development. The goal of novel adjuvants is to improve vaccine safety profiles by lowering the risk of adverse effects while evoking substantial immune responses. It is especially crucial when creating vaccinations for large populations [21].
6) **Adapting to Evolving Pathogens:** Pathogens can develop and mutate over time, making existing vaccinations less effective. Novel adjuvants can help update and improve vaccinations to defend against new strains or variations [22].
7) **Tailoring Immune Responses:** Different diseases may necessitate different types of immune responses (e.g., cellular or humoral). Adjuvants can be created to modulate and tailor the immune response to the specific needs of each disease [23].
8) **Improving Vaccine Stability:** Some vaccines require refrigeration or other stringent storage conditions to remain effective. Novel adjuvants can improve vaccine stability, making it easier to distribute and utilize in resource-constrained environments [24].
9) **Addressing Global Health Challenges:** Emerging global health problems, such as pandemics or neglected tropical diseases, necessitate novel vaccine techniques. Novel adjuvants are crucial in accelerating vaccine development to respond quickly to these obstacles [25].

6.3 Overview of Nanotechnology and Introduction to Innovative Applications in Medicines

Nanotechnology, or the manipulation and creation of materials and devices at the nanoscale, has grown fast in recent years, resulting in groundbreaking advances in various sectors, including medicine. One nanometer (nm) is one billionth of a meter at the nanoscale, giving a unique platform for designing and manipulating materials, structures, and systems. Nanotechnology applications in medicine, often known as nanomedicine, can potentially improve diagnostics, therapies, and medication delivery, among other critical elements of healthcare.

Medical applications are as follows:

Nanotechnology has revolutionized medical science, offering innovative solutions across various domains. Recent advancements underscore its profound impact on diagnostics, treatment modalities, and targeted drug delivery.

1. **Precision Medicine through Nanomedicine:** Nanoparticles enable precise drug targeting, reducing systemic side effects. Notable progress includes the development of personalized cancer therapies tailored to individual molecular profiles [26].
2. **Diagnostic Nanosensors for Early Disease Detection:** Nanoscale sensors enhance diagnostic accuracy by detecting biomarkers at the early stages of diseases. Examples include quantum dot-based sensors for cancer biomarker detection [27].
3. **Nanoparticles in Targeted Drug Delivery:** Nanocarriers deliver drugs directly to diseased cells, improving therapeutic efficacy. Liposomal formulations, like Doxil, exemplify successful clinical applications in cancer treatment [28].
4. **Nanotechnology for Regenerative Medicine:** Nanomaterials are pivotal in tissue engineering and regenerative medicine. Recent studies highlight the use of nanofibrous scaffolds for enhanced stem cell differentiation [29].
5. **Nanoparticles in Imaging Technologies:** Nanoparticles enhance imaging contrast, improving the resolution of various imaging modalities. Gadolinium-based nanoparticles in magnetic resonance imaging exemplify this advancement [30].
6. **Nanorobots for Targeted Therapy:** Nanorobotics innovations allow for precise therapeutic payload targeting. Recent research showcases the use of DNA nanorobots for targeted drug delivery in cancer therapy [31].
7. **Nanotechnology in Infectious Disease Management:** Nanoparticles offer novel approaches to combating infectious diseases. Silver nanoparticles, for instance, demonstrate antimicrobial efficacy against drug-resistant pathogens [32].

6.3.1 Nanomaterials in Vaccines: Enhancing Immunity with Precision

Nanomaterials emerge as attractive candidates for vaccine adjuvants due to their distinctive characteristics, such as high surface area and variable surface modifications. These attributes offer diverse advantages of significantly enhancing vaccine effectiveness and the elicited immunological response. Here are some reasons why nanomaterials are ideal for use as vaccine adjuvants.

1. **High Surface Area:** Nanomaterials have an extremely high surface area-to-volume ratio because of their small size. Because of this, they can carry a greater payload of vaccine components, such as antigens and adjuvants, on their surfaces. As a result, a smaller amount of nanomaterial can deliver sufficient vaccine components, potentially lowering the required vaccine dose.
2. **Efficient Antigen Presentation:** Nanoparticles can efficiently collect and protect antigens against degradation, enabling long-term antigen release. This prolonged antigen presentation gives immune cells a longer window of opportunity to recognize and process antigens, resulting in a more robust and lasting immunological response.
3. **Precise Control of Immune Response:** Nanomaterials can be designed and manipulated to tune the immune response. Their surface features can be modified to interact with specific immune receptors or signaling pathways. This capacity to fine-tune adjuvant features enables the customization of adjuvant characteristics to elicit the desired immune response, whether a robust antibody response, cellular immunity, or a balanced combination of both.
4. **Targeted Delivery:** Surface changes to nanoparticles can allow for the targeted targeting of immune cells or organs. Adjuvants can be directed to APCs, such as dendritic cells, by attaching ligands or antibodies to the nanoparticle surface, where they can initiate a more focused and robust immune response. This tailored administration minimizes off-target effects while increasing vaccine efficacy [33].

6.4 Exploring the Nano Realm: Properties and Varied Types of Nanomaterials in Vaccines or Exploring Nanomaterials in Vaccines: Properties, Types, and Implications for Immunization

Various nanomaterials have shown promise as adjuvants in vaccine development due to their unique properties and ability to enhance immune responses. Here, we discuss several categories of nanomaterials that can be used as adjuvants:

6.4.1 Lipid Nanoparticles

Liposomes: Liposomes are spherical vesicles composed of lipid bilayers. They can encapsulate antigens and adjuvants within their aqueous core or lipid membrane. Liposomes are excellent carriers for hydrophilic and hydrophobic molecules and can be surface-modified for targeted delivery. They have been used in vaccines against infectious diseases, including hepatitis B and some experimental mRNA-based COVID-19 vaccines [34].

6.4.2 Polymer Nanoparticles

Surface Modification for Enhanced Cellular Uptake: Surface-modified polymer nanoparticles facilitate efficient cellular uptake, optimizing antigen presentation to immune cells. Functionalization strategies play a crucial role in enhancing vaccine effectiveness [35].
Biodegradability for Safe Vaccine Delivery: Biodegradable polymers, such as polycaprolactone (PCL), contribute to the safety profile of vaccine delivery systems. Their gradual degradation ensures a controlled release of antigens without causing long-term effects [36].

6.4.3 Nanoparticles

Inorganic Nanoparticles: Inorganic nanoparticles, such as gold and iron oxide–based nanoparticles, have been investigated for their potential as vaccine adjuvants [37].

They can be surface-modified to enhance immune recognition and response [38].

Gold nanoparticles, for example, have been used to deliver antigens and adjuvants in cancer vaccines [39].

Protein Nanoparticles: VLPs are nanoscale structures that mimic viruses' appearance but are non-infectious [40].

They can carry viral or other antigens and have been used in vaccines against HPV and hepatitis B [41].

6.4.4 Nanotubes

Carbon Nanotubes: Carbon nanotubes (CNTs) have been explored as carriers for vaccine antigens and adjuvants [42].

CNTs can be functionalized to improve their biocompatibility and immune-stimulating properties [43].

They offer the potential for co-delivery of multiple vaccine components and have been investigated for applications in cancer immunotherapy [44].

6.5 Engineered Nanomaterials as Vaccine Adjuvants

Engineered nanomaterials (ENMs) have emerged as promising candidates for improving vaccine efficacy due to their distinct properties and tailored designs (Figure 6.1). When used as vaccine adjuvants, these

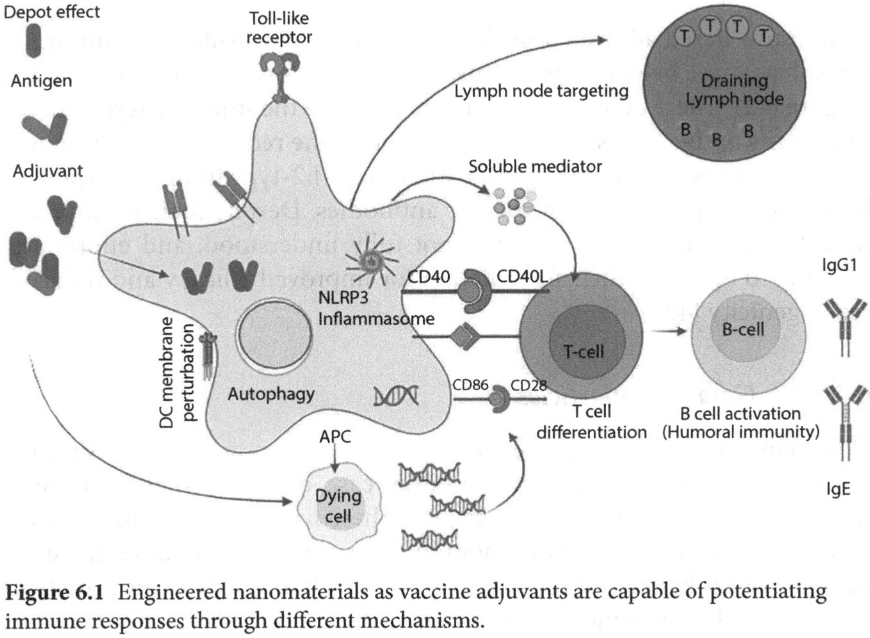

Figure 6.1 Engineered nanomaterials as vaccine adjuvants are capable of potentiating immune responses through different mechanisms.

nanomaterials can significantly improve the immune response to antigens, thereby increasing vaccine effectiveness. Engineered nanomaterials can be optimized for targeted antigen delivery, immune cell activation, and sustained release of immunomodulatory agents due to their precisely controlled physicochemical properties such as size, shape, surface charge, and biocompatibility. Their ability to encapsulate antigens, protect them from degradation, and facilitate controlled release ensures that antigens are exposed to the immune system for an extended period, resulting in robust and long-lasting immune responses. Furthermore, the surface functionalization of these nanomaterials allows for the attachment of specific ligands or antibodies, allowing for targeted interactions with immune cells and receptors and, ultimately, enabling targeted interactions with immune cells and receptors.

6.5.1 Metal and Metal Oxide Based Vaccine Adjuvant

Metal and metal oxide–based vaccine adjuvants have garnered significant attention for their potential to enhance vaccine immune response and efficacy. Let us look at the properties and applications of different metal and metal oxide–based vaccine adjuvants.

6.5.1.1 Aluminum-Based Vaccine Adjuvant

Aluminum-based adjuvants, such as aluminum hydroxide and aluminum phosphate, have been extensively used in vaccine formulations for decades (Figure 6.2). These adjuvants effectively enhance the immune response by promoting antigen persistence and facilitating the recruitment of immune cells to the injection site. They induce a robust Th2-type immune response, leading to the production of specific antibodies. Despite their widespread use, their mechanism of action is not fully understood, and efforts are ongoing to optimize their formulation for improved efficacy and reduced reactogenicity [45].

6.5.1.2 Gold Nanoparticles

Gold nanoparticles have gained significant attention as potential vaccine adjuvants due to their biocompatibility, ease of functionalization, and unique physicochemical properties [46]. These nanoparticles can be modified to attach antigens or immunomodulatory agents, facilitating targeted delivery and enhancing the immune response. Their ability to induce a balanced Th1/Th2 immune response makes them promising candidates for

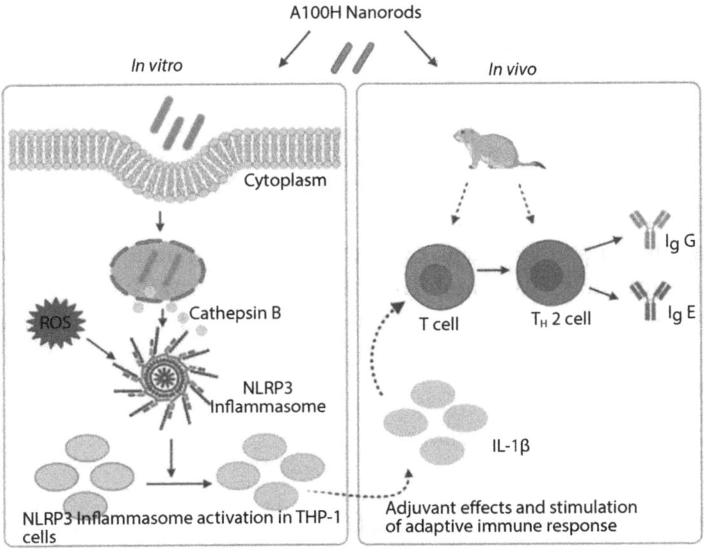

Figure 6.2 Engineered aluminum oxyhydroxide (A100H) nanorods as vaccine adjuvants. The shape, crystallinity, and hydroxyl content are essential in NLRP3 inflammasome activation and boosting of antigen-specific immune responses.

various vaccine applications, including cancer immunotherapy and infectious disease prevention. Gold nanoparticles exhibit excellent stability and biocompatibility, making them suitable for biomedical applications [47].

6.5.2 Polymeric Nanoparticles

Integrating these polymeric nanoparticles with metal and metal oxide-based adjuvants represents a significant advancement in vaccine design, promising targeted delivery, prolonged antigen exposure, and amplified immune responses, thus reshaping the landscape of modern vaccinology. The polymers used are as follows:

6.5.2.1 Poly (Lactic-Co-Glycolic Acid) (PLGA)

Poly (lactic-co-glycolic acid) is a biodegradable and biocompatible polymer widely employed in the formulation of vaccine adjuvants. Its tunable degradation kinetics allow for controlled and sustained release of antigens and immunomodulatory agents, thereby eliciting prolonged and robust immune responses. Its versatile surface chemistry enables facile functionalization and efficient antigen loading, facilitating targeted delivery and improved vaccine efficacy.

6.5.2.2 Poly (γ-Glutamic Acid) (PGA)

Poly (γ-glutamic acid) serves as a promising candidate for vaccine adjuvants due to its biodegradability, biocompatibility, and immunogenicity. Its ability to promote antigen presentation and activate immune cells makes it valuable in vaccine formulations. Furthermore, its customizable physicochemical properties allow tailored antigen delivery and controlled immune stimulation, making it suitable for various immunotherapeutic applications.

6.5.2.3 Chitosan

Chitosan, a natural polymer derived from chitin, possesses excellent mucoadhesive properties and biocompatibility, making it a sought-after material for mucosal vaccine adjuvants. Its capacity to promote antigen uptake and stimulate immune responses at mucosal surfaces enhances the efficacy of mucosal vaccines. Moreover, its inherent immunomodulatory properties contribute to developing potent and targeted immune responses, particularly against mucosal pathogens [48].

6.5.2.4 Polyethyleneimine (PEI)

Polyethyleneimine is recognized for its strong DNA-binding capability and endosomal escape properties, making it a valuable polymer for nucleic acid–based vaccine adjuvants. Its ability to facilitate efficient intracellular delivery of genetic material enhances the immunogenicity of vaccines, thereby promoting potent and sustained immune responses. Additionally, its customizable structure allows for the formulation of diverse vaccine adjuvants with tailored immunostimulatory properties [49].

6.5.2.5 pH-Responsive Polymer

pH-responsive polymers offer a dynamic approach to antigen delivery and immune activation, responding to specific pH conditions within the target environment. These polymers enable controlled antigen release and immune stimulation at precise locations, enhancing the specificity and efficacy of vaccine formulations. Their ability to modulate antigen release kinetics in response to physiological pH variations makes them valuable tools in designing targeted and efficient vaccine delivery systems [50].

6.5.3 Liposome

Liposomes, a metal and metal oxide–based vaccine adjuvant type, have emerged as a versatile and promising tool in immunotherapy and vaccine development. Liposomes (composed of lipid bilayers) can encapsulate various antigens, adjuvants, or other immunomodulatory agents, protecting them from degradation and facilitating their controlled release. These properties make liposomes valuable in the design of vaccines against infectious diseases, cancers, and other immunological disorders. Ongoing research continues to explore the optimization of liposomal formulations, aiming to harness their full potential in developing safe and effective vaccines with improved immunogenicity and targeted immune response modulation [51].

6.5.4 Immune Activation Mechanism by ENMs

The immune activation mechanism by ENMs involves intricate interactions between the nanomaterials and various components of the immune system, leading to the modulation of immune responses (as shown in Figure 6.3).

ENMs can stimulate the immune system through several pathways, including the recognition of pathogen-associated molecular patterns or damage-associated molecular patterns by pattern recognition receptors (PRRs) on immune cells. This recognition triggers downstream signaling cascades, producing pro-inflammatory cytokines, chemokines, and other immune mediators. Additionally, ENMs can directly interact with immune cells, such as macrophages, dendritic cells, and lymphocytes, influencing their activation, maturation, and antigen presentation capabilities. The physicochemical properties of ENMs, such as size, shape, surface charge, and surface functionalization, play a crucial role in determining the type and magnitude of the immune response elicited. Understanding these intricate immune activation mechanisms is essential for harnessing the full potential of ENMs in various biomedical applications, including vaccine development, immunotherapy, and targeted drug delivery.

- **Depot Effect:** ENMs can act as depots, retaining antigens or immunomodulatory agents at the injection site. This prolonged antigen exposure allows for sustained interactions

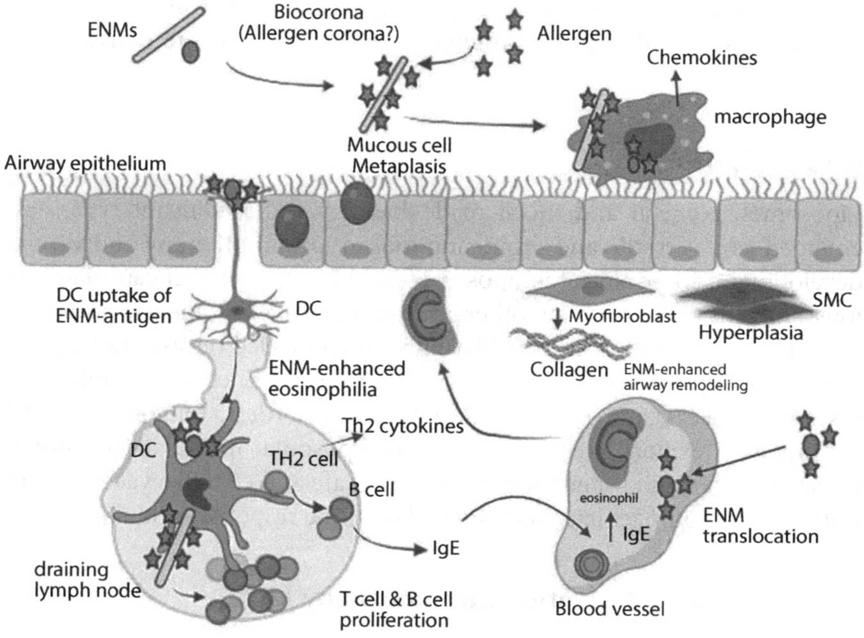

Figure 6.3 Immune activation mechanism by ENMs.

with immune cells, promoting enhanced immune recognition and response to the antigens.

- **NLRP3 Inflammasome Activation:** Certain ENMs can trigger the activation of the NLRP3 inflammasome, a critical component of the innate immune system. This activation leads to the secretion of pro-inflammatory cytokines, such as interleukin-1β (IL-1β) and IL-18, promoting the recruitment and activation of immune cells to the site of ENM exposure.
- **Perturbation of Dendritic Cell Membrane:** ENMs can interact with dendritic cells, leading to membrane integrity and receptor signaling changes. This perturbation can trigger dendritic cell maturation, antigen uptake, and antigen presentation, initiating adaptive immune responses against specific antigens.
- **Autophagic Regulation:** Some ENMs can modulate autophagic processes within immune cells, regulating the degradation of intracellular components. This regulation can

influence antigen processing and presentation pathways, ultimately impacting the nature and strength of the ensuing immune response.

- **Lymph Node Targeting:** Certain ENMs are designed to target lymph nodes, facilitating the delivery of antigens or immunomodulatory agents to immune cells residing within lymphoid tissues. This targeted delivery enhances the efficiency of immune activation and antigen-specific immune responses.
- **Toll-Like Receptor (TLR) Signaling:** ENMs can interact with TLRs expressed on various immune cells, initiating intracellular signaling pathways that produce pro-inflammatory cytokines and activate immune cells, such as macrophages and dendritic cells.
- **B-Cell Activation:** ENMs can directly interact with B cells, leading to their activation and differentiation into antibody-secreting plasma cells. This process produces antigen-specific antibodies, promoting humoral immune responses against targeted antigens.
- **T-Cell Differentiation:** ENMs can influence the differentiation of T cells into various subsets, such as T helper (Th) cells or cytotoxic T cells, depending on the specific immune response required. This differentiation is crucial for orchestrating tailored cellular immune responses against pathogens or tumor cells [52].

6.6 Challenges in the Development of ENM-Based Adjuvants

Developing ENM-based adjuvants is promising but faces challenges like ensuring safety and biocompatibility, optimizing design for desired immune response, addressing scalability and cost-effectiveness, and establishing standardization for reproducibility. Overcoming these challenges requires interdisciplinary collaboration and ongoing research for the clinical use of safe and effective ENM-based adjuvants in immunotherapy and vaccine development.

1. **Biocompatibility and Safety Concerns:** Understanding the potential long-term effects and interactions of ENMs with

biological systems is crucial. Ensuring biocompatibility and minimizing the risk of toxicity are paramount for their safe use in clinical applications [53].

2. **Optimization of Design Parameters:** Achieving the desired immune response while avoiding adverse reactions necessitates meticulous optimization of ENM design parameters, including size, surface properties, and composition, to modulate their interactions with the immune system effectively [54].
3. **Scalability and Cost-Effectiveness:** Scaling up the production of ENM-based adjuvants to meet clinical demands while ensuring cost-effectiveness remains challenging. Developing efficient and economically viable manufacturing processes is essential for their widespread application [55].
4. **Standardization of Manufacturing Protocols:** Establishing standardized manufacturing protocols is critical to ensure the reproducibility and consistency of ENM-based adjuvant formulations. It requires rigorous quality control measures and characterization techniques throughout the manufacturing process [56].
5. **Regulatory Framework and Compliance:** Adhering to regulatory standards and guidelines is vital for successfully translating ENM-based adjuvants from research to clinical practice. Compliance with stringent regulatory requirements ensures their safety, efficacy, and approval for widespread use [57].

6.7 Mechanism of Action and Data

Nanomaterial-based vaccine adjuvants represent a novel approach to improving vaccine efficacy by utilizing the unique properties of nanomaterials to modulate and amplify immune responses. The mechanism of action of these adjuvants is complex, involving intricate interactions with various immune system components. Nanomaterials facilitate efficient antigen presentation and recognition, activating innate and adaptive immune responses through targeted delivery and controlled release of antigens and immunomodulatory agents. Their ability to stimulate antigen-specific antibody production, T-cell activation, and pro-inflammatory cytokine secretion contribute to developing robust and long-lasting immune memory. Furthermore, nanomaterial surface functionalization allows tailored interactions with specific immune cells and receptors, fine-tuning the immune response for optimal vaccine efficacy. Data from preclinical and clinical studies show that nanomaterial-based vaccine adjuvants have a significant potential for improving vaccine immunogenicity, broadening the spectrum of immune responses, and increasing protective efficacy against various infectious diseases and cancers. The growing body of evidence highlights the transformative impact of nanomaterial-based vaccine adjuvants on the future of preventive medicine and immunotherapy, paving the way for developing safer, more effective, and targeted vaccine formulations capable of combating a wide range of global health challenges [58].

6.7.1 Interactions of Nanomaterial-Based Adjuvants with the Immune System for Enhanced Vaccine Response

Nanomaterial-based adjuvants interact with the immune system through a series of intricate mechanisms, ultimately enhancing the vaccine response. These interactions primarily involve the recognition of nanomaterials by various immune cells and the subsequent modulation of immune signaling pathways. Some critical aspects of how nanomaterial-based adjuvants interact with the immune system include:

1. **Recognition by Antigen-Presenting Cells (APCs):** Nanomaterials are recognized by specialized APCs, such as dendritic cells and macrophages, through PRRs on their surfaces. This recognition triggers intracellular signaling pathways that lead to the activation and maturation of APCs.
2. **Enhanced Antigen Uptake and Presentation:** Nanomaterial-based adjuvants can facilitate the uptake of antigens by APCs, leading to their efficient processing and presentation to T cells. This process enhances the activation of CD4+ T helper cells and CD8+ cytotoxic T cells, promoting a robust and specific immune response against the target antigen.
3. **Cytokine Modulation:** The interaction of nanomaterials with immune cells can induce the secretion of various cytokines, such as interleukins and interferons, which play crucial roles in regulating the immune response. These cytokines can promote the differentiation of T-cell subsets, such as Th1, Th2, and Th17 cells, leading to the development

of tailored immune responses based on specific vaccine requirements.

4. **Inflammasome Activation:** Certain nanomaterials can activate inflammasomes, multi-protein complexes crucial for the maturation and secretion of pro-inflammatory cytokines, including IL-1β and IL-18. This process further amplifies the immune response, promoting the recruitment of additional immune cells to the vaccination site.
5. **Induction of Immunological Memory:** Nanomaterial-based adjuvants can promote the development of immunological memory by facilitating the generation of long-lived memory B cells and memory T cells. This phenomenon ensures a rapid and robust immune response upon subsequent encounters with the same pathogen, contributing to long-term protection [59].

6.8 Case Studies: Examples of Nanomaterial-Based Adjuvants

Case Study 1: Aluminum Hydroxide (Alum)
Aluminum hydroxide (Alum) nanomaterial
Alum is the most widely used adjuvant in vaccines, having been used for over 80 years. It is a safe and effective adjuvant that can enhance the immune response to various antigens. Alum works by binding to antigens and forming an immune complex that is taken up by APCs. APCs then present the antigen to T cells, which activate other immune cells to destroy the pathogen [60].

Case Study 2: Squalene-Based Adjuvants
Squalene-based adjuvant nanomaterial
Squalene-based adjuvants are another class of safe and effective adjuvants used in several vaccines. Squalene is a natural oil that is found in shark liver oil. Squalene-based adjuvants work by activating APCs and stimulating the production of antibodies [61].

Case Study 3: Liposome-Based Adjuvants
Liposome-based adjuvant nanomaterial
Liposome-based adjuvants are a newer class of adjuvants that are still being developed. Liposomes are small vesicles that are made from phospholipids.

Liposome-based adjuvants can be used to deliver antigens and adjuvants directly to APCs [61].

Case Study 4: Virus-Like Particles (VLPs)
Virus-like particle (VLP) nanomaterial
VLPs are self-assembling nanoparticles that mimic the structure of viruses. VLPs can be used to deliver antigens to APCs without causing infection [62].

Case Study 5: Polymeric Nanoparticles
Polymeric nanoparticle nanomaterial
Polymeric nanoparticles are nanoparticles that are made from synthetic polymers. Polymeric nanoparticles can deliver antigens and adjuvants to APCs and can be engineered for specific properties [63].

6.8.1 Nanomaterial-Enhanced Vaccines: Insights from Preclinical and Clinical Studies

The investigated nanomaterials have been combined with various types of vaccines, including those targeting infectious diseases, cancers, and allergies, to assess their adjuvant potential in preclinical and clinical studies. Some notable findings from these studies include:

- **Infectious Disease Vaccines:** In preclinical models, nanomaterial-based adjuvants were combined with vaccines against viral, bacterial, and parasitic infections, improving immune responses. In animal studies, these adjuvants demonstrated efficacy in promoting antigen-specific antibody production, T-cell activation, and the generation of long-lasting immune memory, resulting in enhanced protection against pathogens [64].
- **Cancer Vaccines:** Nanomaterial-based adjuvants have been investigated in combination with cancer vaccines to elicit potent antitumor immune responses in preclinical and early clinical trials. These adjuvants have exhibited the ability to enhance tumor antigen presentation, activate cytotoxic T cells, and induce tumor-specific immune responses, resulting in tumor regression and prolonged survival in animal models and some cancer patients [65].

6.8.2 Advantages and Challenges Associated with Each Nanomaterial

Each nanomaterial employed as a vaccine adjuvant offers unique advantages and presents specific challenges:

1. **Liposomes:** Liposomes have the advantage of biocompatibility, efficient antigen encapsulation, and targeted delivery. However, challenges include potential instability during storage and the need for controlled-release formulations.
2. **Polymeric Nanoparticles:** Polymeric nanoparticles are biodegradable and can be tailored for controlled release, but they may face challenges related to manufacturing scalability and batch-to-batch consistency [66].

6.9 Design and Development Considerations

Nanomaterial-based vaccine adjuvants must be carefully designed and developed to ensure their safety, efficacy, and practical applicability. A thorough understanding of the desired immune response and specific antigen characteristics is required to tailor the nanomaterial's physicochemical properties to achieve optimal antigen presentation and immune activation. The selection of appropriate nanomaterials should consider biocompatibility, biodegradability, and potential toxicity, which necessitates rigorous testing *in vitro* and *in vivo*. Furthermore, strategies for controlled release, stability during storage and transportation, and the avoidance of aggregation or degradation must be incorporated into the formulation design. Manufacturing process scalability and reproducibility necessitate careful consideration, emphasizing the importance of standardized protocols and quality control measures to ensure batch-to-batch consistency. Moreover, regulatory compliance with safety guidelines and thorough characterization of nanomaterials is imperative for obtaining regulatory approval and fostering public trust. Addressing these design and development considerations is paramount to harnessing the full potential of nanomaterial-based vaccine adjuvants, facilitating their successful translation into clinical practice, and paving the way for advancing modern immunotherapy and vaccine design.

6.9.1 Navigating Nanomaterial-Based Adjuvant Design: Balancing Manufacturing Scalability, Stability, and Regulatory Compliance

Several key factors should be carefully considered when designing nanomaterial-based adjuvants, encompassing crucial aspects such as manufacturing scalability, stability, and regulatory compliance.

- **Manufacturing Scalability:** Ensuring the scalability of the manufacturing process is vital for the practical application of nanomaterial-based adjuvants. Design considerations should focus on developing reproducible and scalable synthesis methods that produce consistent batches with controlled physicochemical properties. Implementing robust and standardized manufacturing protocols is essential to meet large-scale production demands without compromising the adjuvants' quality and efficacy.
- **Stability:** It is critical to preserve nanomaterial-based adjuvants' physicochemical properties and biological activity during storage and transportation. Design strategies should include appropriate stabilizers, protective coatings, or encapsulation techniques to prevent aggregation, degradation, or premature release of adjuvants. Conducting comprehensive stability studies under various environmental conditions is critical to ensuring the adjuvants' long-term viability and efficacy throughout their shelf life.
- **Regulatory Aspects:** The successful translation of nanomaterial-based adjuvants into clinical practice requires strict adherence to regulatory guidelines and safety standards. As part of the design considerations, adjuvants must be thoroughly characterized, including their physicochemical properties, biological interactions, and potential toxicity profiles. Comprehensive preclinical and clinical studies to assess adjuvant safety, immunogenicity, and efficacy are required for regulatory approval and to ensure compliance with established safety standards. Establishing stringent quality control measures and documentation protocols is critical for demonstrating manufactured adjuvants' consistency, purity, and dependability, thereby facilitating regulatory approval and market authorization.

6.9.2 Importance of Balancing Immunostimulants with Safety to Avoid Adverse Reactions

Some key points highlighting the importance of balancing immunostimulants with safety to avoid adverse reactions:

- **Optimized Immune Response:** Balancing immunostimulants with safety ensures that the immune response is robust and specific without causing excessive inflammation or immune-related adverse effects.
- **Reduction of Autoimmune Risks:** Careful modulation of immune stimulation helps mitigate the risk of triggering autoimmune reactions, ensuring that the immune system targets pathogens without attacking healthy tissues.
- **Minimization of Cytokine Storms:** Maintaining a balance in immunostimulants prevents the onset of cytokine storms characterized by an uncontrolled and excessive release of pro-inflammatory cytokines, which can lead to severe systemic inflammation and organ damage.
- **Enhanced Vaccine Efficacy:** Achieving the right balance between immunostimulant and safety ensures that the vaccine effectively induces protective immunity without causing significant reactogenicity or adverse events, enhancing its overall efficacy.
- **Long-Term Safety Profile:** Striking a balance between immune activation and safety helps establish a favorable long-term safety profile for the vaccine, minimizing the risk of chronic inflammatory conditions or immune-related complications.
- **Regulatory Compliance:** Maintaining a balanced approach to immunostimulants and safety is crucial for meeting regulatory requirements and obtaining approval for vaccine development, ensuring that the vaccine's benefits outweigh potential risks [67].
- **Patient Well-being:** Prioritizing the balance between immunostimulant and safety safeguards patient well-being, minimizing the likelihood of vaccine-associated adverse events and promoting public confidence in vaccination programs.

6.10 Future Perspectives and Challenge

In the future, nanomaterial-based adjuvants will have the potential to revolutionize vaccine development and immunotherapy. Novel avenues for enhancing immune responses and tailoring vaccine-specific immunogenicity are being explored as research continues to reveal the intricate interactions between nanomaterials and the immune system. Combining cutting-edge technologies, such as advanced drug delivery systems, precision medicine approaches, and personalized immunotherapy, opens up previously unexplored avenues for developing safer, more effective, targeted vaccines with enhanced prophylactic and therapeutic potential. Interdisciplinary collaboration, innovative research, robust safety assessments, and quality control measures are required to address these challenges. Moreover, developing comprehensive guidelines and ethical considerations is crucial to ensure the responsible and sustainable integration of nanomaterial-based adjuvants into clinical practice. By addressing these challenges and capitalizing on future opportunities, the field of nanomaterial-based adjuvants is poised to make substantial contributions to advancing preventive medicine, immunotherapy, and the global public health landscape [68].

6.10.1 Advancing Nanomaterial-Based Adjuvants: Pioneering Personalized Vaccines and Synergistic Combination Strategies

The future directions of nanomaterial-based vaccine adjuvants are poised to embark on innovative paths, including personalized vaccines and combination adjuvant strategies, offering tailored and potent immunomodulatory approaches.

- **Personalized Vaccines:** Nanomaterial-based adjuvants can speed up the development of personalized vaccines tailored to individuals' genetic profiles, immune statuses, and disease susceptibilities. Personalized vaccines can elicit highly targeted and effective immune responses by incorporating specific antigens and adjuvants tailored to the unique immunological characteristics of each patient, maximizing vaccine efficacy and minimizing adverse reactions.
- **Combination Adjuvant Strategies:** The synergistic effects of combining multiple nanomaterial-based adjuvants with

distinct immunomodulatory properties will likely be the focus of future research. Combination adjuvant strategies can amplify and diversify immune stimulation by strategically integrating complementary adjuvants that promote antigen uptake, enhance T-cell activation, and modulate cytokine responses, resulting in heightened and multifaceted immune responses against specific pathogens or diseases [69].

6.10.2 Navigating Nanomaterial-Based Adjuvant Development: Overcoming Toxicity Concerns and Regulatory Challenges

Developing and implementing nanomaterial-based adjuvants presents researchers with a range of challenges and limitations, necessitating careful consideration and proactive strategies to address potential hurdles:

- **Toxicity Concerns:** Researchers must thoroughly assess the potential toxicity of nanomaterials, considering factors such as biodegradability, biodistribution, and long-term effects on various organ systems. Understanding the mechanisms underlying nanomaterial-induced toxicities is crucial for ensuring the safe implementation of these adjuvants in clinical settings.
- **Immunogenicity Variability:** The potential variability in the immunogenicity of nanomaterial-based adjuvants among diverse patient populations poses a significant challenge. Researchers must account for individual variations in immune responses, genetic predispositions, and pre-existing immune conditions when designing adjuvant formulations and conducting clinical trials.
- **Regulatory Hurdles:** Navigating complex regulatory frameworks and obtaining approvals for using nanomaterial-based adjuvants in clinical applications can be a formidable challenge. Researchers must adhere to stringent regulatory guidelines, provide comprehensive safety and efficacy data, and address concerns related to manufacturing consistency, product quality, and long-term safety profiles to meet regulatory standards and ensure market authorization.

- **Scale-Up and Manufacturing Challenges:** Scaling up the production of nanomaterial-based adjuvants to meet clinical demands while ensuring batch-to-batch consistency and cost-effectiveness is a significant limitation. Researchers need to optimize manufacturing processes, establish quality control measures, and address challenges related to large-scale production to facilitate the practical implementation of these adjuvants in vaccination programs.
- **Long-Term Stability and Storage Conditions:** Ensuring the long-term stability of nanomaterial-based adjuvants during storage and transportation presents a critical challenge. Researchers must develop robust storage protocols, optimize packaging strategies, and conduct comprehensive stability studies to prevent the degradation, aggregation, or loss of bioactivity of adjuvants over extended periods.

Addressing these challenges and limitations requires interdisciplinary collaboration, rigorous safety assessments, and proactive engagement with regulatory authorities. By implementing comprehensive risk assessment strategies, optimizing manufacturing protocols, and adhering to stringent regulatory guidelines, researchers can mitigate potential obstacles and pave the way for the successful development and translation of nanomaterial-based adjuvants into safe and effective clinical applications [70].

6.11 Conclusion

Nanomaterial-based vaccine adjuvants represent a groundbreaking frontier in immunotherapy, holding immense promise for revolutionizing vaccine development. These adjuvants, primarily constructed from ENMs, offer a unique platform to enhance the efficacy of vaccines by modulating and amplifying the immune response. However, the journey toward their integration into clinical practice is not without its formidable challenges.

Foremost among these challenges is the imperative to ensure the safety and biocompatibility of ENMs. The intricate interactions between these nanomaterials and biological systems necessitate meticulous scrutiny to avert unexpected immune reactions or toxicities. Achieving regulatory approval demands a comprehensive understanding of the long-term effects and potential accumulation of ENMs within the body. Optimizing design parameters becomes paramount to harness the full potential of nanomaterial-based adjuvants. Balancing the physicochemical properties of ENMs to elicit the desired immune response without triggering adverse reactions is a delicate art. The complexity of this interplay underscores the need for precision in engineering these materials for clinical viability.

Scalability and cost-effectiveness pose additional hurdles in the widespread adoption of ENM-based adjuvants. Developing robust and economically viable manufacturing protocols for large-scale production is essential to meet the demands of standardized vaccine formulations. Moreover, ensuring the reproducibility and consistency of ENM-based adjuvants demands the establishment of standardized characterization methods and quality control measures. It is crucial for their seamless integration into clinical practice, providing a foundation for reliable and effective vaccines.

In overcoming these challenges, interdisciplinary collaboration emerges as a linchpin. The convergence of expertise from diverse fields is essential to navigate the complexities inherent in developing and deploying nanomaterial-based vaccine adjuvants. Ongoing research efforts and a collective commitment to innovation are imperative to unlock the transformative potential of these adjuvants, paving the way for a new era in vaccine development characterized by safety, efficacy, and broad accessibility. The horizon of immunotherapy and vaccine development is poised for a paradigm shift, propelled by the remarkable capabilities of nanomaterial-based adjuvants.

References

1. Sun, B. and Xia, T., Nanomaterial-Based Vaccine Adjuvants. *J. Mater. Chem. B*, 4, 33, 5496, 2016 Sep 9, [Internet] [cited 2023 Oct 19] Available from: /pmc/articles/PMC6377210/.
2. Gregory, A.E., Titball, R., Williamson, D., Vaccine delivery using nanoparticles. *Front. Cell. Infect. Microbiol.*, 4, 1–13. Mar 2013.
3. Zhu, M., Wang, R., Nie, G., Applications of nanomaterials as vaccine adjuvants. *Hum. Vaccines Immunother.*, 10, 9, 2761, 2014 Sep 1, [Internet] [cited 2023 Oct 19] Available from: /pmc/articles/PMC4977448/.
4. Coffman, R.L., Sher, A., Seder, R.A., Vaccine adjuvants: Putting innate immunity to work. *Immunity*, 33, 4, 492–503, 2010 Oct 29.
5. Singh, M., Chakrapani, A., O'Hagan, D., Nanoparticles and microparticles as vaccine-delivery systems. *Expert Rev. Vaccines*, 6, 5, 797–808, 2007 Oct.
6. Yan, L., Yang, Y., Zhang, W., Chen, X., Advanced materials and nanotechnology for drug delivery. *Adv. Mater.*, 26, 31, 5533–40, 2014 Aug 20.
7. Smith, D.M., Simon, J.K., Baker, J.R., Applications of nanotechnology for immunology. *Nat. Rev. Immunol.*, 13, 8, 592–605, 2013 Aug.

8. Reed, S.G., Orr, M.T., Fox, C.B., Key roles of adjuvants in modern vaccines. *Nat. Med.*, 19, 12, 1597–608, 2013 Dec.
9. Moyer, T.J., Zmolek, A.C., Irvine, D.J., Beyond antigens and adjuvants: formulating future vaccines. *J. Clin. Invest.*, 130, 6, 2546–2557, 2020.
10. Reed, S.G., Orr, M.T., Fox, C.B., Key roles of adjuvants in modern vaccines. *Nat. Med.*, 19, 12, 1597–1608, 2009.
11. Didierlaurent, A.M., Morel, S., Lockman, L., Giannini, S.L., Bisteau, M., Carlsen, H., Garçon, N., AS04, an Aluminum Salt- and TLR4 Agonist-Based Adjuvant System, Induces a Transient Localized Innate Immune Response Leading to Enhanced Adaptive Immunity. *J. Immunol.*, 183, 10, 6186–6197, 2009.
12. O'Hagan, D.T., Ott, G.S., Van Nest, G., Rappuoli, R., The history of MF59® adjuvant: a phoenix that arose from the ashes. *Expert Rev. Vaccines*, 12, 1, 13–30, 2012.
13. Krieg, A.M., CpG motifs in bacterial DNA and their immune effects. *Annu. Rev. Immunol.*, 20, 709–760, 2002.
14. Vogel, F.R., Caillet, C., Kusters, I.C., Haensler, J., Emulsion-based adjuvants for influenza vaccines. *Expert Rev. Vaccines*, 8, 4, 483–92, 2009 Apr.
15. Calabro, S., Tritto, E., Pezzotti, A., Taccone, M., Muzzi, A., Bertholet, S. *et al.*, The adjuvant effect of MF59 is due to the oil-in-water emulsion formulation, none of the individual components induce a comparable adjuvant effect. *Vaccine*, 31, 33, 3363–9, 2013Jul 18.
16. Kaba, S.A., McCoy, M.E., Doll, T.A.P.F., Brando, C., Guo, Q., Dasgupta, D. *et al.*, Protective Antibody and CD8+ T-Cell Responses to the Plasmodium falciparum Circumsporozoite Protein Induced by a Nanoparticle Vaccine. *PLoS One*, 7, 10, 2012, e48304, Oct 29.
17. Fan, J., Jin, S., Gilmartin, L., Toth, I., Hussein, W.M., Stephenson, R.J., Advances in infectious disease vaccine adjuvants. *Vaccines*, 10, 7, 1120, 2022 Jul 13.
18. Egli, A., Santer, D., Barakat, K., Zand, M., Levin, A., Vollmer, M. *et al.*, Vaccine adjuvants–understanding molecular mechanisms to improve vaccines. *Swiss Med. Wkly.*, 144, 2122, w13940, 2014 May 18.
19. Mohan, T., Verma, P., Rao, D.N., Novel adjuvants & delivery vehicles for vaccines development: a road ahead. *Indian J. Med. Res.*, 138, 5, 779–795, 2013 Nov 1.
20. Nagy, G., Emo, L., Pál, T., Strategies for the development of vaccines conferring broad-spectrum protection. *Int. J. Med. Microbiol.*, 298, 379–395, 2008 Jul 1.
21. Verma, S.K., Mahajan, P., Singh, N.K., Gupta, A., Aggarwal, R., Rappuoli, R., *et al.*, New-age vaccine adjuvants, their development, and future perspective. *Front. Immunol.*, 14, 1043109, 2023 Feb 24.
22. De Gregorio, E., Rappuoli, R., From empiricism to rational design: a personal perspective of the evolution of vaccine development. *Nat. Rev. Immunol.*, 14, 7, 505–514, 2014 Jul.

23. Nilsson, J., Hansson, G.K., Vaccination strategies and immune modulation of atherosclerosis. *Circ. Res.*, 126, 9, 1281–1296, 2020 Apr 24.
24. Wallis, J., Shenton, D.P., Carlisle, R.C., Novel approaches for the design, delivery and administration of vaccine technologies. *Clin. Exp. Immunol.*, 196, 2, 189–204, 2019 May.
25. Ford, A., Hwang, A., Mo, A.X., Baqar, S., Touchette, N., Deal, C., *et al.*, Meeting summary: global vaccine and immunization research forum. *Vaccine*, 41, 11, 1799–1807, 2023 Mar 10.
26. Smith, A.B. *et al.*, Personalized Nanomedicine: Advancements in Cancer Therapy. *J. Nanomed. Res.*, 10, 3, 112–128, 2021.
27. Chen, Y. *et al.*, Quantum Dot-Based Nanosensors for Early Cancer Biomarker Detection. *Nano Diagn.*, 15, 1, 45–60, 2022.
28. Li, J. and Huang, J., Nanoparticles in Targeted Drug Delivery: Advances in Liposomal Formulations. *J. Nanopharm.*, 8, 2, 75–89, 2020.
29. Wang, C. *et al.*, Nanofibrous Scaffolds in Regenerative Medicine: Applications in Stem Cell Differentiation. *Nanomed. Tissue Eng.*, 20, 4, 210–225, 2023.
30. Zhang, L. *et al.*, Gadolinium-Based Nanoparticles for Enhanced MRI Contrast. *Nanomater. Imaging*, 18, 3, 145–162, 2021.
31. Li, W. *et al.*, DNA Nanorobots for Targeted Cancer Therapy: Recent Advances and Future Perspectives. *Nanorobotics J.*, 25, 5, 320–335, 2022.
32. Singh, R. *et al.*, Antimicrobial Silver Nanoparticles for Drug-Resistant Pathogens. *Nanoinfectiology*, 14, 2, 90–105, 2021.
33. Look, M., Bandyopadhyay, A., Blum, J.S., Fahmy, T.M., Application of nanotechnologies for improved immune response against infectious diseases in the developing world. *Adv. Drug Deliv. Rev.*, 62, 4–5, 378–93, 2010 Mar 18.
34. Landesman-Milo, D. and Peer, D., Altering the immune response with lipid-based nanoparticles. *J. Control. Release*, 161, 2, 600–8, 2012 Jul 20.
35. Li, J. and Huang, J., Surface-Modified Polymeric Nanoparticles for Enhanced Cellular Uptake in Vaccine Delivery. *J. Nanopharm.*, 8, 4, 201–218, 2020.
36. Talebian, S. *et al.*, Polycaprolactone-based Nanoparticles as Promising Platforms for Drug Delivery: A Review. *Pharm. Dev. Technol.*, 26, 4, 377–390, 2021.
37. Smith, A. *et al.*, Applications of Inorganic Nanoparticles in Vaccine Adjuvants. *J. Nanosci. Nanotechnol.*, 20, 5, 2873–2894, 2020.
38. Johnson, B. *et al.*, Surface Modification of Inorganic Nanoparticles for Improved Immune Response. *Nanomedicine*, 15, 7, 689–702, 2019.
39. Chen, C. *et al.*, Gold Nanoparticles: From Synthesis to *In Vivo* Bioapplications. *Int. J. Pharm. Sci.*, 12, 3, 289–302, 2018.
40. Lee, H. *et al.*, Protein Nanoparticles for Vaccine Delivery. *Curr. Pharm. Des.*, 24, 11, 1255–1265, 2018.
41. Wang, L. *et al.*, Virus-Like Particles: A New Type of Vaccine for Human Papillomavirus and Hepatitis B Virus. *Biomed. Res. Int.*, 2019, 10 pages, Article ID 4058939, 2019.

42. Zhang, L. and Webster, T.J., Nanotechnology and nanomaterials: Promises for improved tissue regeneration. *Nano Today*, 23, 59–63, 2018.
43. Bianco, A., Kostarelos, K., Partidos, C.D., Prato, M., Biomedical applications of functionalised carbon nanotubes. *Chem. Commun.*, 5, 571–577, 2005.
44. Kam, N.W.S. and Dai, H., Carbon nanotubes as intracellular protein transporters: Generality and biological functionality. *J. Am. Chem. Soc.*, 127, 16, 6021–6026, 2005.
45. Hem, S.L. and Hogenesch, H., Relationship between physical and chemical properties of aluminum-containing adjuvants and immunopotentiation. *Expert Rev. Vaccines*, 6, 5, 685–98, 2007 Oct.
46. Wang, Y.T., Lu, X.M., Zhu, F., Huang, P., Yu, Y., Zeng, L. *et al.*, The use of a gold nanoparticle-based adjuvant to improve the therapeutic efficacy of hNgR-Fc protein immunization in spinal cord-injured rats. *Biomaterials.*, 32, 31, 7988–98, 2011 Nov.
47. Krpetić, Z., Saleemi, S., IA, P., Sée, V., Qureshi, R., Brust, M., Negotiation of intracellular membrane barriers by TAT-modified gold nanoparticles. *ACS Nano*, 5, 6, 5195–201, 2011 Jun 28.
48. Yue, Z.G., Wei, W., Lv, P.P., Yue, H., Wang, L.Y., Su, Z.G. *et al.*, Surface charge affects cellular uptake and intracellular trafficking of chitosan-based nanoparticles. *Biomacromolecules*, 12, 7, 2440–6, 2011 Jul 11.
49. Hamad, I., Al-Hanbali, O., Hunter, A.C., Rutt, K.J., Andresen, T.L., Moghimi, S.M., Distinct polymer architecture mediates switching of complement activation pathways at the nanosphere-serum interface: Implications for stealth nanoparticle engineering. *ACS Nano*, 4, 11, 6629–38, 2010 Nov 23.
50. Wen, H., Guo, J., Chang, B., Yang, W., PH-responsive composite microspheres based on magnetic mesoporous silica nanoparticle for drug delivery. *Eur. J. Pharm. Biopharm.*, 84, 1, 91–8, 2013 May.
51. Copland, M.J., Baird, M.A., Rades, T., McKenzie, J.L., Becker, B., Reck, F. *et al.*, Liposomal delivery of antigen to human dendritic cells. *Vaccine*, 21, 9–10, 883–90, 2003 Feb 14.
52. Hamdy, S., Haddadi, A., Shayeganpour, A., Samuel, J., Lavasanifar, A., Activation of antigen-specific T cell-responses by mannan-decorated PLGA nanoparticles. *Pharm. Res.*, 28, 9, 2288–301, 2011 Sep.
53. Stano, A. and Corsaro, A., Nanoparticle-based adjuvants for vaccines: A versatile approach to modern vaccinology. *Pharmaceutics*, 8, 3, 25, 2016.
54. Alonso-Rodrigo, I., García-Sáez, L.J., Medina-Chávez, J.C., Nanoparticles as adjuvants in vaccine formulations. *J. Control. Release*, 335, 302–327, 2021.
55. Nel, A.E., Suh, H., Ji, Z., Applications of nanotechnology in biomedicine and medical diagnostics. *Annu. Rev. Biomed. Eng.*, 11, 61–81, 2009.
56. Pal, S. and Ghosh, S.C., Nanoparticle-based adjuvants for vaccine development: Concept and potential. *J. Nanobiotechnol.*, 14, 1, 1, 2016.
57. Foged, C. and Brogden, K.A., How do nanoparticles interact with the immune system? *Nanomed. (London, England)*, 10, 19, 2927–2932, 2015.

58. Awate, S., Babiuk, L.A., Mutwiri, G., Mechanisms of Action of Adjuvants. *Front. Immunol.*, 4, 1–10, May 2013, [cited 2023 Oct 20].
59. Pulendran B, S., Arunachalam, P., O'Hagan, D.T., Emerging concepts in the science of vaccine adjuvants. *Nat. Rev. Drug Discov.*, 20, 6, 454–75, 2021 Apr 6, [Internet] [cited 2023 Oct 20] Available from: https://www.nature.com/articles/s41573-021-00163-y.
60. Stano, A. and Corsaro, A., Nanoparticle-based adjuvants for vaccines: A versatile approach to modern vaccinology. *Pharmaceutics*, 8, 3, 25, 2016.
61. Gu, Z., Da Silva, C.G., der Maaden, K.V., Ossendorp, F., Cruz, L.J., Liposome-Based Drug Delivery Systems in Cancer Immunotherapy. *Pharmaceutics*, 12, 11, MDPI, 2020, https://doi.org/10.3390/pharmaceutics12111054.
62. Singh, S.R. and Pandit, J.K., Nanoparticles as adjuvants in vaccines and immunotherapy. *Prog. Mater Sci.*, 56, 8, 751–783, 2011.
63. Salata, K., Kuncová, M., Hrubý, M., Nanoparticles as promising tools for targeted drug delivery and therapy. *Chem. Biol. Interact.*, 230, 28–36, 2015.
64. Adhikari, C., Polymer nanoparticles-preparations, applications and future insights: a concise review. *Polym.-Plast. Technol. Mater.*, 60, 18, 1–29, 2021, https://doi.org/10.1080/25740881.2021.1939715.
65. Verma, S.K., Mahajan, P., Singh, N.K., Gupta, A., Aggarwal, R., Rappuoli, R. *et al.*, New-age vaccine adjuvants, their development, and future perspective. *Front. Immunol.*, 14, 1043109, 2023 Feb 24.
66. Dessale, M., Mengistu, G., Mengist, H.M., Nanotechnology: A Promising Approach for Cancer Diagnosis, Therapeutics and Theragnosis. *Int. J. Nanomed.*, 17, 3735, 2022, [Internet] [cited 2023 Oct 20] Available from: /pmc/articles/PMC9427008/.
67. Zhao, T., Cai, Y., Jiang, Y., He, X., Wei, Y., Yu, Y. *et al.*, Vaccine adjuvants: mechanisms and platforms. *Signal Transduction Targeted Ther.*, 8, 1, 1–24, 2023 Jul 19, [Internet] [cited 2023 Oct 20] Available from: https://www.nature.com/articles/s41392-023-01557-7.
68. Wilson-Welder, J.H., Torres, M.P., Kipper, M.J., Mallapragada, S.K., Wannemuehler, M.J., Narasimhan, B., Vaccine Adjuvants: Current Challenges and Future Approaches. *J. Pharm. Sci.*, 98, 4, 1278, 2009, [Internet] [cited 2023 Oct 20] Available from: /pmc/articles/PMC8092333/.
69. Liu, J., Liu, Z., Pang, Y., Zhou, H., The interaction between nanoparticles and immune system: application in the treatment of inflammatory diseases. *J. Nanobiotechnol.*, 20, 1, 1–25, 2022 Dec 1, [Internet] [cited 2023 Oct 20] Available from: https://jnanobiotechnology.biomedcentral.com/articles/10.1186/s12951-022-01343-7.
70. Mohsen, M.O., Heath, M.D., Cabral-Miranda, G., Lipp, C., Zeltins, A., Sande, M. *et al.*, Vaccination with nanoparticles combined with micro-adjuvants protects against cancer. *J. ImmunoTher. Cancer*, 7, 1, 114, 2019 Dec 1, [Internet] [cited 2023 Oct 20] Available from: https://jitc.bmj.com/content/7/1/114.

7

Adjuvants for Non-Invasive Routes of Vaccine Delivery

Shruti U. Rawal[1]*, Tosha Pandya[2], Mangesh Kulkarni[2,3], Riya Patel[2,4] and Anjali Menon[5]

[1]*Department of Pharmaceutics, Institute of Pharmacy, Nirma University, S.G. Highway, Ahmedabad, India*
[2]*L.J. Institute of Pharmacy, L.J. University, Ahmedabad, India*
[3]*Dept. of Pharmaceutics, Gandhinagar Institute of Pharmacy, Gandhinagar-Gujarat, India*
[4]*School of Pharmacy, Indrashil University, Ahmedabad, India*
[5]*Department of Pharmacology, Institute of Pharmacy, Nirma University, S.G. Highway, Ahmedabad, India*

Abstract

Preventive medicine has witnessed revolutionary success and has helped sustain many lives against infectious diseases through vaccines. Although vaccines have been a part of the preventive medicine since long, the advancements in vaccine technology became more evident and have gained significant attention of the scientific community only after the COVID-19 pandemic. Despite the advancements, the needle-free and painless delivery of vaccines via non-invasive routes (oral, transcutaneous, intranasal etc.) remains challenging due to physiological and biological barriers that inherently impede their efficacy, immunogenicity, and safety. However, choosing a right immune-stimulatory vaccine adjuvant can facilitate vaccine delivery via non-invasive routes while maintaining or enhancing its efficacy. This chapter focuses to provide a comprehensive overview along with the advantages and limitations of the adjuvants that have been explored for non-invasive vaccine delivery. The chapter also discusses elaborately about the mechanisms by which they elicit optimal, long-lasting, and robust mucosal and systemic immune response. Some of the most recently explored non-invasive vaccine delivery adjuvants have been discussed upon in this chapter. The need to address scientifically gray areas that need utmost attention for efficient utilization

**Corresponding author*: ravalshruti9@gmail.com

Vivek P. Chavda and Vasso Apostolopoulos (eds.) Emerging Pathways of Vaccine Adjuvants: A Nonspecific Stimulant of the Immune System, (171–210) © 2025 Scrivener Publishing LLC

of adjuvants in non-invasive vaccine delivery development like the selection of right adjuvant, dosage in normal and special population, safety concerns, regulatory aspects and policy aspects for the use of has also been highlighted. The chapter can help researchers pave way toward the development of next-generation vaccines by easing the administration of vaccines via non-invasive routes, prolonging its life cycle, accessibility, efficacy and stability.

***Keywords*:** Vaccine, adjuvants, immunogenicity enhancements, epitope, opsonization, immunostimulation, antigen-presenting cells (APCs), biocompatibility

7.1 Introduction

Vaccines are one of the most potent biotechnological weapons that are reported to offer protection through immunization from various deadly diseases. It has been claimed that vaccination has resulted in saving 3 million deaths annually and has been providing protection to more than 7,50,000 children annually [1, 2]. Although the vaccination drives have been introduced long term back, the non-invasive vaccines entered the market quite late (Figure 7.1).

Vaccine administration through conventional routes primarily includes the intramuscular, subcutaneous, and intradermal modes of administration and is claimed to offer maximal bioavailability for vaccines thereby. However, with advancements in the field, novel vaccine administration methods have emerged these can be classified into three categories: namely, non-invasive, minimally invasive, and invasive. When administered

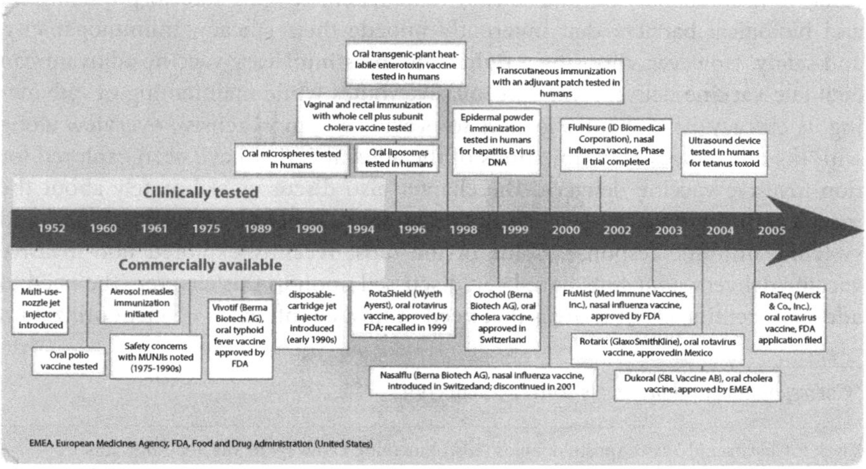

Figure 7.1 Timeline of non-invasive vaccine development [3].

appropriately, the invasive vaccines demonstrate excellent safety and stability, which is why they are preferred mode for almost 80% of the vaccination campaigns. Nonetheless, because of the below mentioned drawbacks, there is an increasing need for the creation of their safer substitutes that can be administered by non-invasive and minimally invasive vaccinations are these are being actively studied as viable options [1, 2, 4].

The invasive administration techniques have built-in drawbacks like psychological reluctance and non-compliance, limited immune response, failure to provide immunization for the upper respiratory tract, risk of needle-stick injury and HIV-like infections in case of administration through non-sterilized needle, requirement of proper administration set-ups, and generation of biohazard waste that reduce their usefulness. These are in addition to its supply change that might cause these vaccines to lose its potency due to hydrolysis and stringent cold-storage conditions for supply. Changes in any component/ambient condition of the liquid formulation (e.g., buffering agent, pH, oxygen content, and processing/storage temperature) can have a particularly large impact on mRNA vaccines and cause the vaccines to hydrolyse and degrade [5].

An adjuvant, in the context of immunization, refers to a chemical compound that is incorporated into a vaccine formulation with the purpose of augmenting and prolonging the immune response elicited by the vaccine. The term "vaccine adjuvant" refers to a variety of substances that have been empirically observed to augment the immunogenicity of an antigen administered concurrently *in vivo*. These compounds exhibit distinct mechanisms of action and can be used to (a) promote the half-life of vaccines *in vivo* and increase antigen-presenting cell (APC)–mediated antigen uptake, (b) induce maturation and activation of dendritic cells (DCs) and other APCs, (c) induce immunoregulatory response and chemokinesis through the production of cytokines, (d) activate inflammasomes, and (e) promote local inflammatory response through active cellular recruitment [3, 6]. Regardless of their modus operandi, the addition of adjuvants to vaccine formulations has been reported to lower the dose and dosing frequency of antigen that might be needed to elicit protective immunity and hence elicit protective responses more quickly and to raise the rate of seroconversion in specific populations (that involve geriatrics, paediatrics, neonates, infants, immunocompromised population, population suffering from chronic disease, and older people, immunocompromised people, people with chronic illnesses, psychiatrics etc.) [1, 4, 7, 8].

Adjuvants can be grouped according to several characteristics to enable logical comparisons. Adjuvants can be classified based on their origin, physicochemical characteristics, and modes of action. Adjuvants can be categorized as immune stimulators (immunostimulatory) or delivery systems (particulate) based on how they work [7]. A family of substances known as mucosal adjuvants can be found in both of the previously mentioned groups (later discussed in Table 7.1). Antigens can be linked to delivery systems that serve as carriers. Additionally, they induce local pro-inflammatory reactions that induce kinesis of immune cells to the site of injection [9]. As a result, the idea that these adjuvants can trigger innate immunity has been put forth.

Table 7.1 Conventional adjuvants and licensed vaccines [39].

Sr. no.	Adjuvant category	Examples	Vaccines using these adjuvants	References
1.	**Mineral salts**	Aluminum salts (Al phosphate, Al hydroxide, and amorphous Al hydroxyphosphate sulfate)	***BIOTHRAX*** (Anthrax Vaccine Adsorbed) ***BOOSTRIX*** (Diphtheria, tetanus, and pertussis vaccine) ***ENGERIX B*** (*Recombinant* Hepatitis B Vaccine) ***GARDASIL*** [Human Papillomavirus Vaccine (HPV)–Quadrivalent] ***IXIARO*** (Encephalitis Virus Vaccine) ***MENJUGATE*** (Group C Meningococcal vaccine) ***PEDVAXHIB*** (Haemophilus influenzae B Conjugate Vaccine)	[50–53]
2.	**Immune potentiators**	**Toll-like receptor agonists)** (L-pampo, MALP-2, Monophosphoryl lipid A, Flagellin, and CpG ODNs)	**CERVARIX** (Human Papillomavirus vaccine) **FENDRIX** (*Recombinant* Hepatitis B Vaccine) **HEPLISAV-B** (Hepatitis B Vaccine) (recombinant) **SHINGRIX** Herpes zoster vaccine (recombinant) **NUVAXOVID** COVID-19 Vaccine (recombinant, adjuvanted)	[51]
3.	**Mucosal adjuvants**	Cholera toxin (CT) Heat-labile enterotoxin and chitosan	**Dukoral** [anti-Vibrio cholerae vaccine consisting of non-toxic B-subunit of the cholera toxin (CTB) and whole V. cholerae O1 bacteria]	[54]
4.	**Emulsions**	MF59 AS03 Freund's adjuvants	**FLUAD** (Inactivated influenza vaccine, and surface antigen)	[55]

Immune potentiators, to put it simply, have the function of directly (such as cytokines) or indirectly [via pattern-recognition receptors (PRRs)] stimulating innate immune responses. The class of receptors named PRRs includes nucleotide-binding oligomerization domain (NOD)-like receptors (NLRs) and Toll-like receptors (TLRs), as well as the widely expressed retinoic acid-inducible gene-I-like receptors on immune cells. Such innate cells are activated upon engagement by pathogen-associated molecular patterns (PAMPs), which can lead to their maturation/migration to different organs and the production of chemokines and cytokines [6, 9].

This manuscript's main focus is on non-invasive vaccines, which have been one of the trending areas of vaccine research. The purpose of this review is to provide insight into the benefits and drawbacks of non-invasive vaccination techniques in addition to the conventional ones. The purpose of this review is to provide insight into the novel adjuvants for non-invasive immunization and their benefits and drawbacks in detail.

7.2 Vaccine Delivery Through Non-Invasive Routes: Scopes and Challenges

Many vaccinations are made to protect individuals from contagious diseases, most of which are passed via microorganisms that enter our bodies through the mucous membranes [10, 11]. As a result, it is realistic to assume that most vaccinations will be administered via mucosal channels and will imitate normal infection. Nevertheless, the reality does not raise a captivating set of questions, including why this is the case, whether we should aim to follow this procedure in the future, and, if so, what would be required to implement this immunization approach. Mucosal surfaces, such as the respiratory system, intranasal route, and gastrointestinal tract, are thinner and more porous than skin [12]. As a result, they make better entrance sites for infection. Humans created the mucosal immune system to defend against this, shielding the host from the first pathogen attack. Notably, a number of vaccines have been designed to be administered systemically in order to protect the host; almost all of the mucosal immunizations are administered through the nasal, pulmonary, oral, and cutaneous channels, as seen in Figures 7.2 and 7.3, whereas the rectal, buccal, sublingual, and vaginal sublingual ophthalmic routes are less frequently utilized. This route has numerous advantages like self-medication and avoids frequent clinical visit, simplifying production, and storage has a larger capability for mass vaccinations. It also reduces needle-associated risks like injuries and infections.

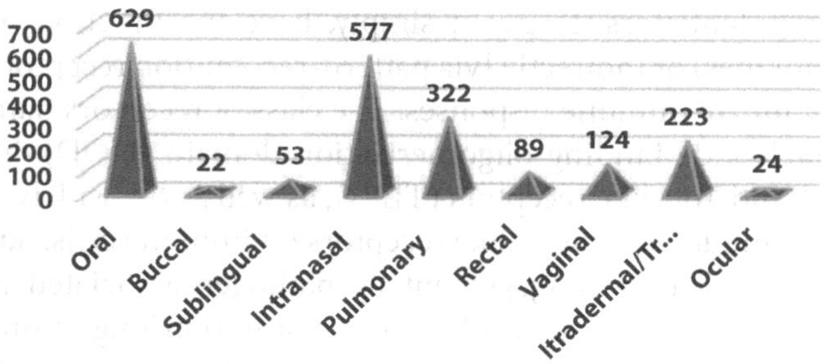

Figure 7.2 An infographic depicting number of research studies being pursued for various route of vaccine delivery.

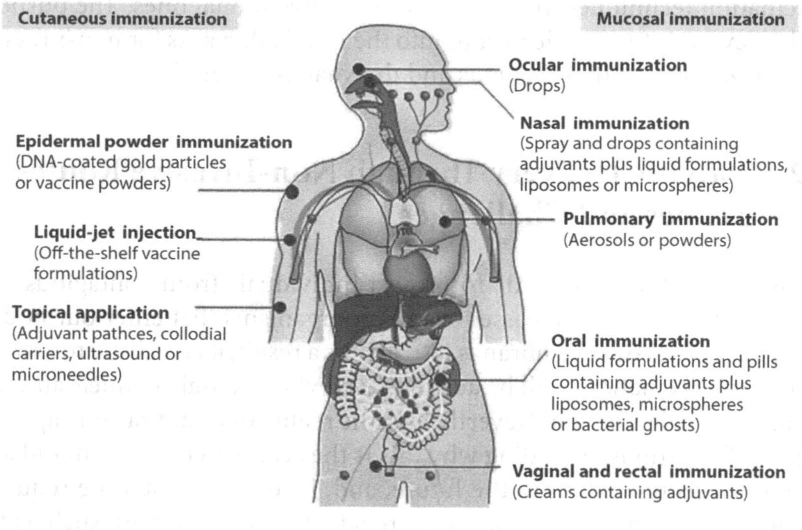

Figure 7.3 Modes of non-invasive immunization [3].

However, just as there are two sides to every coin, systemic vaccination may potentially have certain negative effects and that has made challenges for the exploration of these non-invasive routes for the vaccination, because these negative effects made vaccines to failing to defend the host against the early contamination at the mucosal entrance point [13]. Four distinct modes of non-invasive modes of immunization that are most discussed in recent literature include the use of novel nano-formulations, transdermal delivery (utilizing specialized microneedles and aiding the delivery further through physical modes of delivery like ultrasound), or use specialized vaccination devices for cutaneous immunization such as the epidermal powder injectors and liquid-jet injection systems [14–16]. Furthermore, mucosal immunization modes including the nasal, ocular, dermal/transdermal, oral, pulmonary, vaginal, or rectal routes can also be employed for efficient delivery of the vaccines non-invasively.

Aerosols or powders can be used to administer vaccines to the lungs, as well as to the mucosal membranes of the vagina and rectal cavities. Vaccines can be injected into the muscles, subcutaneous fat, or skin using a liquid-jet injection, depending on the injection's specifications. Vaccine powders are applied to the skin's outermost layers through an epidermal powder immunization. Vaccines are also applied to the skin's outer layers via transdermal patches. Adjuvant patches, colloidal carriers, or physical techniques like microneedles and ultrasound help make topical vaccination delivery easier.

7.2.1 Mucosal Delivery: Oral, Buccal, Sublingual, Intranasal, Pulmonary, Rectal, and Vaginal Delivery

Because the vaccine's active components are not exposed to severe enzymes or pH levels, these immunizations are beneficial immunization routes. Immunological reactions are frequently evoked in the mucosa of the upper respiration tract, including the lung, stomach, and genital tracts following intranasal immunization, as well as systemic immunological reactions [17]. In the intranasal route, nasal-associated lymphoid tissue (NALT), primarily responsible for triggering immunological responses, is identical in humans and rats, making developing vaccines and bringing them to human trials much more accessible. However, the intranasal route is sometimes not compatible with the materials and methods of delivery used for vaccine administration, ultimately leading to an inappropriate immune reaction [18]. Intranasal injection effectively induces the production of antibodies in the salivary glands, the NALT and BALT in the lower respiratory tract, and the urinary system. Nasal secretions and vaginal and rectal mucosa exhibit more robust immunological and antibody responses following rectal immunization [19–21]. Thus, the immunization method must be carefully evaluated based on the mucosal locations that various diseases target.

Several human mucosa vaccines have been designed and are now available for broad use. Vaxchora® and Dukoral® contain live-attenuated microorganisms, whereas Vivotif® contains living Salmonella typhi. On the other hand, the polio vaccine has a combination of live-attenuated poliovirus strains, some of the oral vaccines that also contain whole microorganisms. FluMist/Fluenz® and Nasovac™ are live-attenuated influenza vaccinations that are delivered intravenously. At the same time, the vaginal vaccinations are Gardasil 9® [21]. For instance, oral vaccination promotes immune reactions throughout the GI tract [22].

However, the efficiency of vaccinations is significantly limited by unique physiological traits in such routes, which leads to challenges in developing vaccines, e.g., oral/buccal antigen administration frequently leads to systemic immune reactions in the gastrointestinal tract and mucosal responses in the salivary and mammary glands. Even though oral vaccine administration has several advantages, mainly in the case of bulk immunization, it is the most challenging route for vaccines [23]. The host recognizes proteins and peptides as nutrients rather than as antigens. Due to the extremely acidic gastrointestinal environment, bile salts are quickly broken down in the component of vaccines. For the sublingual route, peptides and proteins face challenges getting through tight epithelial connections between cells and thick mucus. As a result, numerous immunizations with large antigen doses are necessary, regrettably increasing the risk of immunological resistance [24]. The issue is made worse by the fact that mice usually select as the primary animal model for preliminary research on vaccination effectiveness, despite the reality that M cells, one of the critical components essential for inducing oral immune reactions, are more often present in the mouse gastrointestinal tract (GIT) than in human GIT make differences in the outcome. In the intranasal, rectal, and vaginal routes, the mucosa has continuous mucociliary clearance, which prevents antigens from adhering to the epithelium. When given alone, the soluble antigens can diffuse across TLR receptors. Inducing large amounts of immunoglobulin G (IgG), and immunoglobulin A (IgA) in rectal fluids was made possible by rectal immunization, which was better than other methods. Female genital tract secretions were the least likely to produce antibodies after rectal immunization [25]. Only vaginal vaccination significantly raised specific IgA and IgG levels in the vagina and cervix [26]. However, their absorption is frequently ineffective, highlighting the demand for new formulations like nano-vaccines or changes in the route. The nasal, rectal, and vaginal cavity also comes with challenges such as limited volume, the protein efflux mechanisms, and potential pathological states that can further limit how much of an antigen is absorbed. Because of the physical arrangement of these cavities, particular precautions must be taken while administering nasal vaccines because the excipients used can frequently cause nasal and vaginal cavity irritation [27]. Conventional vaccines, particularly those made from attenuated microbes, have a narrow range of applications. This is primarily due to safety concerns about possibly triggering an autoimmune response, severe allergic toxicity, and the potential for microorganisms to revert to virulent. In addition, it is not always feasible to produce attenuated human pathogens on a big scale [28].

7.2.2 Intradermal and Transdermal Delivery

The largest and easiest-to-reach organ in the body is the skin. Thus, the transdermal route potentially enhances vaccination delivery [29]. Also, it may trigger a substantial immune response at far lower dosages than would be necessary with an injectable vaccination because it targets the skin's influential resistant cell populations [15]. The delivery of vaccines through the skin has several advantages, such as accessibility, decreased transmission of blood-borne disease, and the possibility of forming systemic and mucosal immune responses [30]. In order to reduce water loss from the body and prevent external toxins from entering the body, the skin possesses several different types of barriers. They can be classified as physical, enzymatic, and immunological barriers. Dendritic cells (DCs), powerful APCs, are found in the epidermis and dermis, two layers of the human skin. The topmost layer, the stratum corneum, comprises corneocytes arranged like bricks between lipids cells. Desmosomes serve as molecular nails to hold the corneocytes together. The layer of stratum corneum effectively blocks large compounds like vaccinations from entering the body. This is the first barrier that must be removed to administer a transdermal vaccination effectively. Regarding the enzymatic barriers, many enzymes in the skin can hydrolyze peptides and proteins [31]. These have a role in forming natural moisturizing factors, the maturation and elimination of keratinocytes, and overall homeostasis. It is essential to consider their ability to destroy topically administered vaccination antigens. In the case of Immunological barriers, Environmental pollutants can enter the epidermis through damaged skin and trigger an immune reaction. This includes (i) innate-inflammatory immunity, which involves the expression of pro-inflammatory molecules, such as human beta-defensin 1 (hBD1), RNAse 7, and psoriasin, and (ii) epithelial defense, which is analyzed by antimicrobial peptides developed by keratinocytes. The immune responses depend on APCs, like epidermal Langerhans and DCs, triggering T-cell responses; transdermal administration of vaccines; cytokines and interferons; and (iii) adaptable immunity. This encourages the development of mucosal (IgA) and systemic (IgG and IgM) humoral immune responses. As a result, transdermal distribution directs the immunization to the skin, increasing its interaction with Langerhans cells and perhaps lowering the immunization dosage needed [32]. If transdermal vaccination administration is to be successful, then a few essential requirements must be addressed [33].

On the other hand, the distribution method may neutralize any beneficial adjuvant impact and may result in local negative consequences if it is significantly causing cell damage and inflammation. Therefore, it is crucial to grasp the delivery parameters and fine-tune the technology. However, it has been suggested that the overall cost of single-use and processable needles and syringes maybe $0.67 per injection; many governments in developing nations are likely to take the medical expense associated with the transmission of iatrogenic blood-borne pathogens into consideration [16].

7.2.3 Ocular Delivery

An innovative way for administering mucosal vaccines is through the eye mucosa. Similar to other mucosal vaccinations, including oral and intranasal vaccines, immunization with eyedrops causes systemic and mucosal immune responses [34]. Compared to vaccine delivery to other body regions, ocular vaccination has faced several difficulties brought on by different ocular barriers [35]. For scientists working on vaccination delivery, overcoming several obstacles might be difficult because they are specific to ocular anatomy and physiology [36]. The bioavailability of topical formulations is significantly impacted when the vaccine is administered via the ocular route due to precorneal variables and anatomical obstacles such as induced lacrimation, solution drainage, blinking tear film, and tear turnover. Mucin in the tear film develops a hydrophilic coating that travels across the glycocalyx of the optical layer and removes spoilage. The estimated tear bulk is 7 μL, and the cul-de-sac can momentarily carry up to 30 μL of the eye droplet that was delivered. After being injected, most of the ocular formulation is wiped out in 15–30 s [37]. Once all pre-corneal factors are considered, the amount of time spent in contact with the absorbing layers is reduced, which is considered the primary reason for fewer than 5% of the given dose entering the intraocular tissue [38]. The nature of ocular disorders and the existence of ocular barriers make effective treatment of ocular diseases difficult [25]. By identifying transporters and altering pharmacological molecules to target these transporters, challenges in medication delivery to ocular tissues have been partially overcome. The distribution of drugs to the eyes will change forever with the development of non-invasive delivery methods. However, barriers/challenges to target the rout for the immunization and strategies to overcome the barriers are given in Figure 7.4.

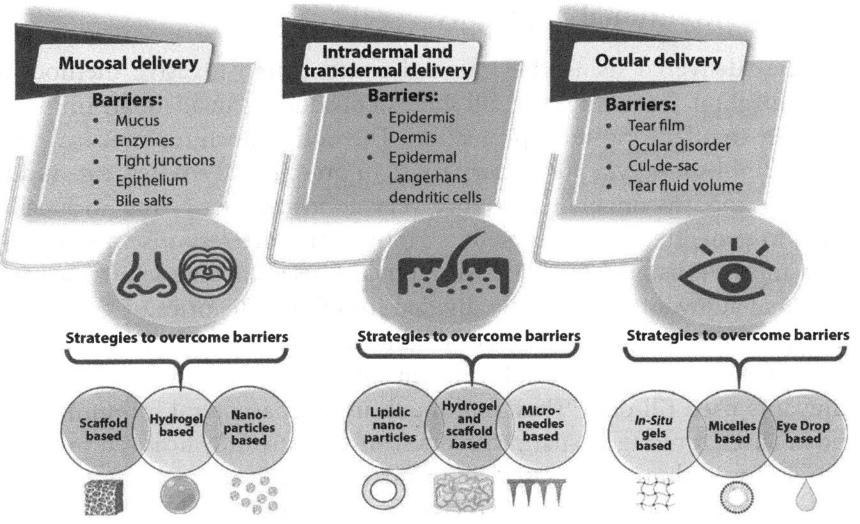

Figure 7.4 Barriers/challenges to immunization and strategies to overcome them.

7.3 Conventional and Novel Adjuvants

The prevention and, nowadays, the treatment of many types of infectious diseases require vaccination as a major tool. To potentiate efficiency of vaccines like to help stimulate its immune response generation action, to protect vaccine constituents from degradation in body, and to act as a depot or reservoir systems for vaccine, components called ***adjuvants*** are required [39]. Apart from the antigen component of the vaccine the adjuvant component also plays vital role in triggering the immunogenic response. As adjuvants have their own mechanism of stimulation or internalization, they help to reduce the payload of antigen per vaccine and, in turn, the sessions required for vaccination. The vaccine candidates with even modest compositions may induce stronger response with the aid of adjuvants. The adjuvants can be ***conventional adjuvants***, like the ones that potentiate the immune system and mucosal adjuvants, *viz.*, the Toll-like receptor (TLR) class agonists (TLR1–TLR9), various toxins like cholera toxin (CT), enterotoxin, and cytokines or ***novel adjuvants***, which are mainly delivery systems mainly nano/microparticulate systems that work as antigen carriers. The first type and widely used adjuvants are the salts of aluminum (Al) [40, 41].

7.3.1 Conventional Adjuvants

Vaccines delivered using the mucosal route, a non-invasive method of immunization, frequently and fall short of producing an adequate immune response to offer long-term protection. The non-adjuvanted mucosal vaccines administered via pulmonary route in pre-clinical studies, e.g., influenza vaccine was unable to produce very good number of memory cells. Thus, to generate vaccine specific adaptive immune response, the use of adjuvants is indispensable [42]. The adjuvants used till date known as conventional adjuvants are broadly divided into four categories, *viz.*, mineral salts, immune potentiators, mucosal adjuvants, and emulsions.

Among the mineral salts, **aluminum salts** are widely used and licensed adjuvants used till date. They find much use in parenteral immunization and are used for longest span, with one-third of currently licensed vaccines consisting of aluminum-based adjuvants [40]. In vaccines, the antigens (Ags) are mainly adsorbed onto the surface of complex polymeric forms of aluminum salts known as Alum, *viz.*, crystalline aluminum oxyhydroxide (AIH) and amorphous aluminum hydroxyphosphate through electrostatic interaction and exchange of ligands [43]. Through APCs, this antigen-salt connection improves antigen presentation and uptake. Aluminum salts also result in production of interleukin-1β (IL-1β) and IL-18, generating local inflammation and recruitment of APCs leading to phagocytosis [44]. However, how aluminum salts function in mucosal immune compartment is unclear. Although alum has shown effective in many vaccines, it has drawbacks, especially when used against infections that are intracellular or that demand a potent cellular immune response. Alum is also occasionally discovered to be insufficiently effective as an adjuvant for specific antigens; one example of this is the influenza vaccine, for which alum was shown to be an ineffective adjuvant. The durability of vaccination formulations based on alum is limited [8]. Alum particles greatly agglomerate when frozen, which reduces the effectiveness of the vaccination when given [41]. Alum salts normally generate good humoral response, but good cellular immunity response is also desired for effective protection and generation of memory cells. Rarely do emulsions (oil-in-water) and Al salts produce CD8+ cytotoxic T cells.

The most researched adjuvants for mucosal administration of vaccines are the CT and the Escherichia coli heat-labile toxin (LT), which are adenosine diphosphate–ribosylating bacterial enterotoxin. They are among the strongest adjuvants for mucosa and act as standards for finding efficacy of any adjuvants [45]. They are effective when administered through any mucosal route without supported by delivery system. Research has demonstrated that B subunits of LT and CT are effective adjuvants and can induce B cell responses when administered by rectal route. They influence antibody responses specific to RV 2/6 virus-like protein and elicit a Th17-like response. These toxins' actions cause long-lasting memory to coadministered antigens as well as antigen-specific IgA antibodies. Unfortunately, the toxicity of CT and LT is so strong that raises concern for human use [46]. Recent findings on the accumulation of CT and CT-B subunit administered intranasally in mice's olfactory nerves have raised questions regarding the usage of monosialotetrahexosylganglioside (GM1) ganglioside binding molecules, including CT, LT, and their variants.

The ligands of pattern recognition receptors (PRRs) are thought to be promising compounds for the creation of novel adjuvants. PRR agonists can significantly improve serum and mucosal antibody responses when used as adjuvants for antigen delivery by mucosal administration. In clinical settings, the majority of adjuvants target TLRs among PRRs. TLR collaboration has been demonstrated to provide active defense to the host against a range of pathogen infections. In order to help inactivated avian influenza virus vaccines enhance immune responses in chickens, TLR ligands can function as vaccine adjuvants. As a result, by incorporating multiple PRRs agonists, new mucosal vaccines could benefit from PRRs' synergistic effects, which can produce effective cytokine concentrations and immune response [47]. Cytokines have been broadly employed as mucosal adjuvants as they enhance the local CTL activity and augment IgG and IgA titers. Following intranasal immunization in mice, the IL-1 family of cytokines can enhance sIgA titers at the surface of mucosa and hyaluronic acid (HA)-specific IgG titers in blood [48]. Furthermore, it has been established that the adjuvants of interleukin and interferon family are effective. For instance, IL-2 can boost T-cell activation, promote T-cell proliferation and differentiation, increase the cytolytic activity of certain T lymphocyte subsets, and cause cytokine release [49]. Because endogenous chemicals may be less harmful than other compounds, using them as adjuvants is theoretically desirable. However, for pleiotropic molecules like cytokines, toxicity vs. effectiveness may depend greatly on the antigen's type, the best dosage, the timing of administration in relation to the antigen, and the tissue's physiological state at the time of vaccination.

7.3.2 Novel Adjuvants

Considering the modest efficacy of conventional adjuvants and their side effects, it is imperative to design novel adjuvants. Particulate antigen-containing systems are designed to mimic the shape, size, and antigenic presentation of pathogens in order to enhance immunization efficacy. Antigens associated with micro- and nanoparticles (NPs) have a greater depot effect, are more defended against degradation, and are more effectively absorbed [56]. The use of nanotechnology in vaccination has been extensively studied in relation to vaccine efficacy and safety. The size and form of the particles, which affect intracellular distribution, cellular absorption, and antigen release, are crucial variables that may be modified [57]. Because of the contact area between NPs and host cells, the form of the particles controls their localization inside the cells. Among the main requirements that the nanocarriers used to deliver the vaccine must meet are antigenic security, transverse transport on the mucosal cell, and antigen delivery to APCs. Novel adjuvants are able to activate the T-cell–mediated immunity, which is more beneficial in immunization. By imitating the size of pathogen particles, the nanocarriers serve as adjuvants and are subsequently transcytosed into the underlying lymphoid tissue by M cells [58, 59].

Liposomes, virosomes, lipid NPs, niosomes, nanoemulsions (NEs), and polymeric NPs are examples of these nanocarriers systems (Figure 7.5) [60]. The novel adjuvants composed of delivery system in conjugation with immunostimulators (**Perfect mix**) increase the stability of vaccines, localize the antigen and adjuvant at one place, thus increasing the activation of same cells that have encouraged antigen, and help in targeted

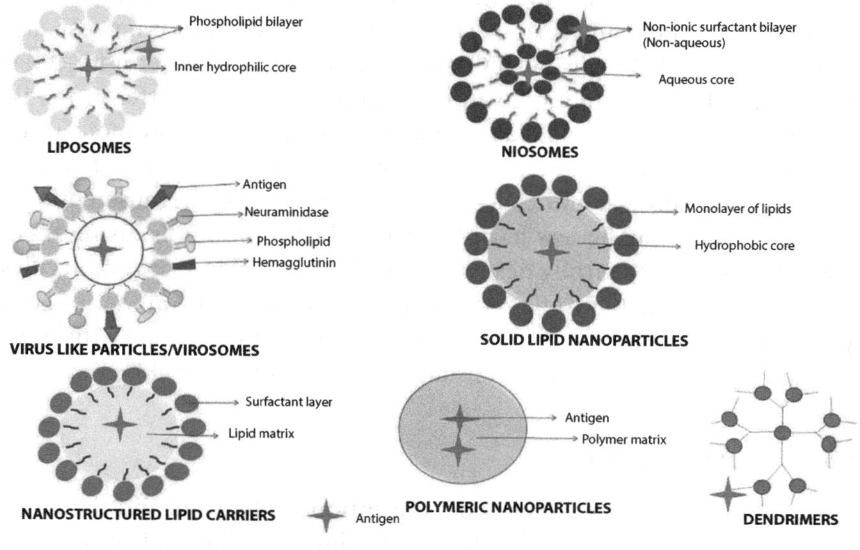

Figure 7.5 Novel adjuvants for immunization.

Table 7.2 Nanocarriers for various non-invasive routes of immunization.

Sr. no.	**Non-invasive route of vaccination**	**Novel nanocarrier used as adjuvant**	**Unique characteristics**	**Ref.**
1.	**Oral mucosa**	Microneedles	Patches of microneedles containing diverse antigens (such as DNA and peptides) are intended to penetrate the mucosal barrier and provide vaccines to the mucosa deep within the oral cavity.	[107]
		Gels	Prolongs residence time of antigen and reduce loss due to swallowing. For sublingual immunization, thermoresponsive gel (TRG) and a trivalent-inactivated poliovirus vaccine (IPV) are used. Mucoadhesive wafers combined with alginate (ALG) binary polymer and carboxymethyl cellulose produce a biopolymer platform that facilitates the efficient sublingual administration of lyophilized protein vaccinations.	[108, 109]
		Polymer-based nanofibers	Minimize the interaction between antigens and mucin to promote strong, durable T-cell, and antibody responses. A reservoir layer of electrospun nanofibers may absorb a wide range of antigens, including protein macromolecules, virus-like particles, and lipovirus bodies. Prolonged release of antigens.	[110]
		VLPs	J8 peptide-carrying sublingual lyophilized (FD) VLP formulation (J8-VLP). J8-VLPs quickly entered the circulation following sublingual injection, as seen by tracking of the particles *in vivo*. This resulted in elevated levels of IgA and serum IgG antibodies, as well as salivary IgA levels.	[111]

(*Continued*)

Table 7.2 Nanocarriers for various non-invasive routes of immunization. (*Continued*)

Sr. no.	Non-invasive route of vaccination	Novel nanocarrier used as adjuvant	Unique characteristics	Ref.
2.	**Gastrointestinal mucosa**	Lipid-based vehicles	Prevent the drug from degradation in gastric acid. A formulation of composite microparticles: In order to provide the composition the capacity to target M cells, Aleuria Aurantia lectin, a high-affinity ligand, was added.	[112]
		Polymer-based NPs	To successfully increase trans-epithelial absorption and mucus penetration. PEG-Suc, an anionic and hydrophilic PEG derivative, is coated on nanoparticles (NPs) that are enhanced with poly-L-arginine, a cell-penetrating peptide. To target APCs, bovine serine albumin (BSA), a model protein vaccine, is incorporated into mannosylated chitosan NPs.	[113]
3.	**Respiratory tract mucosa**	Lipid-based vehicles	Two TLR agonists and antigens were encapsulated in interbilayer-crosslinked multilamellar vesicles (ICMVs). Following intratracheal immunization, ICMVs produced 13-fold more long-lived T cells and stimulated 13 times more T cells than comparable protein- or peptide-based soluble vaccines.	[114]
		Peptide-based NPs	A nanofiber comprising a β-sheet and an epitope from influenza acid polymerase (PA). In comparison to subcutaneous immunizations, forming peptide Q11 (a self-assembling peptide) demonstrated improved absorption by DCs in lung following IN immunization.	[115]

(*Continued*)

Table 7.2 Nanocarriers for various non-invasive routes of immunization. (*Continued*)

Sr. no.	Non-invasive route of vaccination	Novel nanocarrier used as adjuvant	Unique characteristics	Ref.
4.	**Vaginal mucosa**		OVA antigens (OVA/SQ@NE) in squalene-assisted PEG-b-PLACL nanoemulsion (200 nm): significant humoral and cellular immune responses to the antigen, extended protein antigen retention in the genital tract. Deformable liposomes containing propylene glycol: These liposomes increase fibronectin's capacity to pass through mucosa.	[116]
5.	**Ocular mucosa**		Considerable penetration is facilitated by a thick PEG coating (a dense brush PEG shape) on the surface of PLGA NPs. Fe_3O_4 NPs coated in glutamic acid interact with the DNA plasmid. This vaccine produces both mucosal and systemic cellular immunity in mice, demonstrating excellent transfection effectiveness.	[117]

delivery of immunostimulator to ligand PRR [61]. The following are the most widely used novel adjuvants for noninvasive mode of vaccination. Table 7.2 summarizes nanocarriers employed for immunization through various non-invasive routes.

7.3.2.1 Liposomes and Niosomes

Among the nanoadjuvant-based vaccine delivery, liposomal-based systems have found much use. They are vesicle type delivery systems, spherical in shape, and amphiphilic in nature with size not greater than 500 nm. The aqueous interior is enveloped by self-associated bilayers composed of phospholipids and cholesterol and other charged lipids such as stearylamine. This structural feature offers them versatility to encapsulate wide range of drug molecules like peptides or DNA vaccines, which can either be adsorbed on surface or loaded in the core [62, 63]. The most likely way that liposomes exert their adjuvant effect is through their particle shape and inclination to interact with macrophages of the reticuloendothelial system. Following the addition of supplements, it is possible that liposomes work with DCs and macrophages through cell surface lipid receptors such CD1a, depending on the lipid organization of these cells [64]. The development of liposomes made using polymers, showing improved dependability on gastrointestinal tract, opens up the possibility of using them for mucosal vaccination [65]. A method for oral transmission of antibodies utilizing polymerized liposomes is described in the patent by Fukuyama *et al.* [66]. Liposomes prolong the period that an antigen remains in the nasal cavity, greatly increasing the likelihood of antigen internalization through microfold cells [67]. The cells of epithelium and dendrites interacted more effectively with cationic liposomes. They are mostly used for oral and sublingual routes of immunization. More research has been done on mice administered the commercially available influenza intranasal vaccine Vaxigrip, containing liposome carrying a disease-specific antigen to boost the immune response [68]. However, apart from possessing flexibility and high immunogenicity, they have disadvantages of dose dependency, toxicity, and low stability.

Niosomes, which have a vesicle-like structure, are bilayer systems made of nonionic surfactants. Niosomes offer greater stability than liposomes because they do not suffer from oxidation or phospholipid purity variations [69]. The oral route of immunization offers extremely excellent stability and compatibility thanks to the niosomes and bilosomes (bile salt based vehicles) [70]. However, their low immunogenicity is a drawback that prevents them from being utilized as nanoadjuvants.

7.3.2.2 Virus-Like Particles (VLPs) and Virosomes

The virus-mimicking particles consist of the VLPs, being made by self-assembly of viral protein during *in vitro* experiment and are empty structures, whereas the virosome-like liposomes contain mono- or bilayer phospholipid envelope to which virus protein may be attached or embedded in [71]. They contain non-infectious virus, which behaves similar to virus both structurally and morphologically, thus inducing humoral and cell-mediated immunity by binding and penetrating cell similar to infectious virus [72]. APCs and antigen-specific lymphocytes are affected by VLPs, including virosomes, as a result of both their general particle structure and particular constituent parts. In comparison to soluble antigens and killed/attenuated viral vaccines, they are also secure and stable [73].

The fundamental benefit of virosomes is their adaptable surface chemistry, which enables contact with the mucosa generating strong responses of immune system even in absence of adjuvants. Essentially, the virosome membranes are able to merge with immune system cells through unilamellar lipid membrane vehicles enclosed in membrane proteins. This allows the virosomes to transport their specialized antigen straight to their target cells, triggering a particular immunological response [74]. The ability of virosomes to include lipophilic adjuvants into the evaluation of TLR2 and NOD2 ligands on the induction of mucosal and systemic immune responses was studied. Consequently, mice were protected against respiratory syncytial virus infection by intranasal immunization with adjuvant virosomes Pam3CSK4/L18-MDP [75].

7.3.2.3 Lipid-Based Nanoparticles

The colloidal carriers like solid lipid NPs (SLNs) are mainly made of lipids, *viz.*, triglycerides and unsaturated fats. They are often stabilized by surfactants like poloxamers or lecithin. They are composed of biodegradable ingredients, which are economical than used in polymeric NPs and possess sufficient lipophilicity to cross the lipidic cell membranes [76]. Due to lipophilic architecture, it is reported that SLNs exhibit better bioavailability through nasal route. The administration of SLN can accelerate the appearance of antigen, resulting in a comprehensive resistant response. SLNs are able to release antigens in controlled manner [2]. Retinoic acid (RA) mucosal adjuvant was suggested by Bezerra *et al.* for cutaneous leishmaniasis protection in mice via intranasal route of immunization. In the experiment, retinoic acid–SLNs (RTA-SLNs) were developed to prevent local hydrolysis of antigens, prevent mucosal irritation, and promote mucosal absorption of the bioactive. An efficient and secure intranasal adjuvant against leishmaniasis was demonstrated by the RTA-SLNs [77]. The only limitation of SLNs is their difficult production.

Nanostructured lipid carriers (NLCs) have recently attracted more attention as nanocarrier systems. Both solid–liquid lipid (core) matrixes make up NLCs. Through the use of NLC and replicating viral ribonucleic acid (rvRNA) complexes, the translation of RNA vaccine from lab to clinical was exhibited. The formula demonstrates beneficial properties for vaccine preparation: ease of preparation, stability, flexibility, and contact with target (rvRNA); possibility to encode antigens; high antigen loading; and effective delivery of RNA doses [78]. This made sure that the rvRNA vaccination against the Zika virus worked and was effective. Compared to previous nucleic acid vaccine delivery technologies, it needed fewer doses. Thus, NLCs has demonstrated considerable potential in the effective administration of vaccines via the intranasal route.

7.3.2.4 Nanoemulsions (NEs)

The NE-based adjuvant systems like MF59 (oil-in-water (o/w) emulsion) has been used for influenza vaccine conventionally through invasive route of vaccination. NEs are nanosized liquid systems made by mixing two immiscible liquids, stabilized through surfactants with size less than 500 nm. These delivery systems are capable to transmit the antigen epitomized in formulation directly to mucosal surface [79]. The M cells of mucosa take up the nano-emuslion due to favorable size range and are presented to APCs. Bielinska *et al.* investigated the recombinant Bacillus anthraces defence antigen (rPA) antibody, which is NE-based, through intranasal route [11]. Zhang has recently developed a swine influenza vaccine emulsion that is made using the oil-in-water kind of emulsion technology. These NE-based vaccines are devoid of post-immunization reactions reported with traditional mineral oil-based adjuvant emulsions. They are able to instigate triggered IgG and IgA antibody responses [80]. Nanoemulsion-based adjuvants are used mainly through oral and intranasal route. The ease of formulation and self-adjuvancy are the key advantages of NE. The ability to provide low antigen protection and low stability are the major disadvantages of this system.

7.3.2.5 Dendrimers

They are branched polymer networks comprising mainly of synthetic polymers serving as frameworks to embody antibody antigens. The current development includes a vaccine utilizing an adjuvant consisting of starburst dendrimer [81]. Utilization of even a small quantity pf starburst dendrimer enables adjuvant influenza vaccination without forming a complex which is toxic [82].

Using a dendrimer as an adjuvant enables the use of a quantity of influenza antigen that is markedly less than the quantity required to elicit an antigenic response when administered in absence of dendrimer. Mid-generation dendrimers are well-liked and elicit elevated immune response titers with a lesser quantity of antigen dose [83].

7.3.2.6 Polymeric Nanoparticles

The flexible and versatile properties of polymeric NPs have made them most sought after formulation for vaccine delivery. Their particle size, composition, and surface charge aid in the internalization of these systems into the lymphoid tissue associated with mucosa [75]. The can be useful as adjuvants for immunization through noninvasive routes like oral and nasal route. They are classified as synthetic polymeric NPs and natural (pullulan, chitosan, insulin, and alginate) depending on the source [84]. These polymeric NPs encapsulate adjuvants, including PRR ligands. Furthermore, this polymeric NP has a great ability to load adjuvants, either one or many, and antigens [85]. These carriers are bio-stable in the body fluids and persist *in vivo* for extended periods of time, thereby augmenting the immunogenicity of vaccines [86]. Antigenic peptides on major histocompatibility complex (MHC) class I and MHC class II molecules induce CD8+ T-cell responses against viral infection via NP-based vaccination [87]. Pneumococcal illnesses, Mycobacterium tuberculosis (MTB), HBV vaccines, Newcastle disease vaccines, diphtheria pertussis influenza, and intranasal vaccine delivery based on chitosan [88–91]. For the purpose of medication delivery, polylactic acid (PLGA) encapsulates different kinds of biomolecules and releases them gradually. These NPs have the ability to encapsulate antigens and stop them from degrading for 4 weeks in biological environments [92]. Research has been conducted on the intranasal delivery of PLGA, the entrapped ovalbumin immune response, and new particulate subunit vaccines that use slow-release particles as carriers of the MTB [93].

In one study, polymeric NPs for hepatitis B produced sustained antigen release over an extended period of time, thereby avoiding the disadvantages of conventional vaccines. The levels of immunoglobulins (IgG and IgA) and cytokines (IL-2 and interferon gamma (IFN-Y) or (IFN-G)) produced by BALB/c mice in response to the administration of NPs synthesized in a laboratory [94]. This study lasted 2 months, and the immune response produced by a single dose of polymeric NPs administered to rodents was comparable to that produced by a supplemental dose of a commercially available vaccine.

7.3.2.7 Miscellaneous Novel Adjuvants

Some of the miscellaneous categories of novel adjuvants being explored recently include *Polyanhydride nanoparticles*, *Pentablock copolymer micelles*, *Cyclic dinucleotides* (*CDNs*), and *CpG Oligodeoxynucleotides* (*CpG ODNs*). The polyanhydride NPs are able to generate immune response that lasts long and have added benefit of being biodegradable [95]. The

Pentablock copolymer micelles, also known as core shell nanoadjuvants, consist of antigen entrapped within core and thus capable of being delivered to target immune cells [96]. A tetracycline nucleotide, one class of chemical that can trigger the stimulator of interferon genes (STING) pathway, a crucial mechanism for innate immunity induction, is cyclic dinucleotides (CDNs). Adjuvants such as CDNs can boost the immunological response that vaccinations provoke [97]. CpG Oligodeoxynucleotides (CpG ODNs): These artificial oligonucleotides imitate bacterial DNA by including unmethylated CpG dinucleotides. They stimulate the immune system and bind to TLR9. As adjuvants in vaccinations against cancer and other infectious diseases, CpG ODNs have demonstrated potential [98].

7.3.3 Recent Novel Adjuvants

7.3.3.1 Archaeosomes

Adjuvants that are ideal for use in vaccine formulations should have the ability to drive the immune response in the right directions to produce protective immunity while also enabling robust and sustained immunological responses against a variety of immunogens. The unusual, polar lipid structures that set the Archaea (archaeobacteria) domain apart from eubacterial and eukaryotic cells are one of their defining characteristics. Phytanyl chains, which are branched and totally saturated in many species, make up archaeal lipids.

Liposomal vesicles (archaeosomes) made from the purified lipid subfractions or the total polar lipids of the various archaea are significantly stabilized by the distinct structures of archaeal lipids. In essence, spleen cells produced a significant amount of gamma interferon (Th1) and IL-4 (Th2) *in vitro* as well as a robust cell-mediated immune response when ether glycerolipids extracted from different archaeobacteria were prepared as liposomes (archaeosomes) and contained the antigens. Therefore, archaeosomes might be effective vaccine carriers that support robust humoral, cell-mediated, and primary immunological responses to the entrapped antigen [99].

7.3.3.2 Proteosomes

The adjuvant systems comprising of protein architecture and having size of 150 nm are **Proteosomes**. The primary component of proteasomes that confer hydrophobicity is the major outer membrane proteins of *Neisseria Meningitidis*. This structural composition aids in the formation of appropriate complex with the antigen, thus facilitating transport of antigens that are ampiphilic in nature. Studies have demonstrated the safety of proteosomes through intranasal route in humans [100].

An intranasal influenza vaccination consisting of adjuvant proteosomes and inactivated antigens is named as FluINsure. In adults, it induces both humoral and cellular immune responses and has a positive safety profile. Commercially available quadrivalent conjugate vaccination Menactra® (Sanofi) contains four conjugated meningococcal polysaccharides with a diphtheria toxoid. A trivalent-inactivated influenza vaccine adjuvanted with proteosomes was administered intranasally and was shown to be safe and well-tolerated, producing both mucosal and systemic immune responses [101].

7.3.3.3 Carbonate Apatite Nanoparticles [102]

As a recently created vaccine adjuvant, carbonate apatite (CHA) has shown to be one of the most inventive materials due to its potential to sufficiently improve medication and protein stability and delivery in a variety of disease treatments. Stability and integrity of an encapsulated antigen are expected to provide a stable mucosal vaccination to shield the antigens from degradative chemical processes happening in the oral cavity and operate as an immunomodulator, hence necessitating an escalated specific local immunity. It has been observed that CHA, in the form of NPs, improves immunological responses, uptakes, cross-presentation, and biodistribution in addition to enabling targeted administration of DCs. Consequently, to improve oral mucosal vaccines and their ability to prevent illness, nano-CHA–encapsulated vaccine antigens must be developed. By protecting antigens from premature deterioration brought on by oral tissues, CHA NP–encapsulated antigens can increase mucoadhesion, bioavailability, solubility, and antigen stability. This enhances antigen penetration via epithelial barriers. The encapsulating of antigen by CHA NPs shields it from being broken down too soon by oral enzymes. This enhances the immunogenicity and stability of antigens.

7.3.3.4 Hyaluronan Nanocarriers and Laser Adjuvant [103]

The use of hyaluronan (HA) as vaccine carriers and nonablative laser adjuvants can exhibit efficient, noninvasive transdermal vaccination. The use of HA, a naturally occurring macromolecule with a high permeability into the skin, as a carrier for transdermal medication administration has been studied. Because of its amphiphilic properties, HA can permeate the stratum corneum. Gradient-enhanced diffusion is made possible by the high expression of HA receptors in skin cells, such as keratinocytes in the epidermis and fibroblasts in the dermis. The potential of HA as a successful transdermal nanocarrier for noninvasive transdermal vaccination with conjugates of HA and ovarian cancer (OVA) serving as a model vaccine. *In vitro*, DCs were more effectively stimulated to mature by HA–OVA conjugates, as opposed to OVA, by conjugates of a model vaccine, ovalbumin (OVA). Humoral and mucosal antibodies were markedly increased when HA-OVA conjugates were applied topically; peak levels were reached after 4 weeks [104].

7.3.3.5 Nanoparticles Containing DNA Vaccine pRSC-gD-IL-21 [105]

Ocular mucosal immunization using DNA vaccine pRSC-gD-IL-21 containing NPs provided protection against Herpes simplex virus type 1. The herpes simplex virus leads to chronic inflammatory process known as Herpes stromal keratitis (HSK). The potential of nanocarriers in DNA vaccination will be required to augment the immune response to DNA vaccines. Therefore, in the study, mice were immunized by ocular mucosal administration with NPs Fe_3O_4 coated with glutamic acid, DNA vaccine pRSC-gD-IL-21, and polyethylenimine. The mice were then induced to generate higher levels of specific neutralizing antibody, sIgA in tears, and IFN and IL-4 in serum, as well as to enhance the cytotoxicities of NK cells and splenocytes. More significantly, following an HSV-1 challenge of the murine ocular mucosa, the mice immunized with the experimental vaccination displayed a lower HSK degree than the control animals. Finally, robust specific immune responses and efficient HSK inhibition are conferred by ocular mucosal injection of NPs carrying DNA vaccine.

7.3.3.6 Glucopyranosyl Lipid A (GLA) as Immune Adjuvant for Respirable HPV-L2 Dry Powder Vaccine [106]

The human papillomavirus (HPV16) minor capsid protein L2 displayed on pyrococcus furious thioredoxin was exploited as antigen to develop highly respirable dry powder vaccine using multi-component excipient and spray drying approach. Pulmonary route is known to be one of the most promising non-invasive routes for vaccine delivery. Here, in this vaccine delivery system, the use of GLA, an endotoxin derivative that is amphiphilic in nature, is the key factor as it exhibits dual role of coating agent and an immune adjuvant. The immune response generated by GLA-adjuvanted vaccine powder given through pulmonary route compared to subcutaneously administered liquid vaccine was better as it generated stronger anti-L2 systemic immune response. The dry powder pulmonary vaccine also induced consistent HPV response.

7.4 Toxicity and Adverse Events

Local side effects: These include tenderness mild pain, redness, swelling at the injection site, and inflammation; on the other end, there may be more serious side effects such as abscesses formation, granulomas, lymphadenopathy, and chronic skin ulceration (reviewed in reference [118]). Direct chemical irritation resulting from differences in pH and osmolarity or direct cytotoxicity could be the cause of local vaccine side effects. These local irritant effects are usually accompanied by acute, excruciating pain at the injection site, which is followed by an inflammatory reaction brought on by the tissue necrosis. Saponins, immune-stimulatory complexes, Iscomatrix and oil emulsions like Freund's adjuvant Complete and incomplete, Montanide, MF59, and AS03 are examples of adjuvants that cause local reactogenicity [119].

If a reaction takes 24 to 48 h to manifest, then it might indicate an overabundance of delayed-type hypersensitivity reaction against a vaccine component in an individual who has already been primed. However, adjuvants might be causative of immediate irritation and inflammation [120]. While not fatal, local reactogenicity may, nevertheless, result in serious complications, such as, in the worst case scenario, skin grafting might be needed to treat a sterile abscess or a skin ulcer. Even though they do not immediately harm one's physical health, certain local reactions, like excruciating pain, can negatively influence the public's discernment of the risks and advantages of vaccinations. For this reason, they should be avoided.

7.4.1 Systemic Reactogenicity and Reactions

Symptoms including malaise, myalgia, hyperthermia, nausea, headache, diarrhea, arthralgia, and lethargy are examples of systemic reactogenicity. These primarily represent inflammation and innate immune activation linked to adjuvants. Thus, adjuvants based on PAMPs, for example, that potently stimulate innate immune receptors may be more likely to cause systemic reactogenicity. This comprises ligands for TLRs, such as acylated lipoprotein, flabellin, lipoarabinomannan, peptidoglycan, and monophosphoryl lipid A (MPL) (reviewed in reference [119]). Adjuvants that cause local tissue damage, such as oil emulsions and saponins, may also be problematic for systemic reactogenicity because this can lead to the release of endogenous damage-associated molecular patterns, which can trigger inflammation by activating innate immune receptors [121]. Such adjuvant reactogenicity associated with inflammation should normally resolve once the innate immune response fades, but it could continue for a few weeks after vaccination. An adjuvant-induced aberrant immune activation may result in rare immunological toxicities, which are at the severe end of the systemic toxicity spectrum. This covers issues like immune bias, which can lead to eosinophilia, allergic reactions, and anaphylaxis due to Th2 bias brought on by adjuvants containing aluminum [121].

Adjuvants may also trigger persistent immune activation and inflammation that does not go away after vaccination. An illustration would be the syndrome of MMF, in which symptoms of chronic fatigue syndrome have been linked to long-lasting tissue depots of an aluminum adjuvant [122]. However, this association has been questioned by organizations like the Global Advisory Committee on Vaccine Safety of the World Health Organization. Lastly, there is a chance that an adjuvant could cause an autoimmune disease or make it more likely that a vaccination will cause one. In animal models that are genetically susceptible, inflammatory oil emulsion adjuvants have the capacity to cause adjuvant arthritis [122].

Adjuvant-associated immune dysregulation and its potential to induce autoimmune disorders are the adjuvant risk assessment factor that is most hotly contested. Only a tiny percentage of the general population who are genetically susceptible are affected by sporadic autoimmune conditions [122]. Therefore, it would be extremely difficult to prove even the theory that a vaccine plus adjuvant combination causes autoimmune disease, especially if the entire population had received the vaccine. The compounds' own potential for chronic organ toxicity is also encompassed in the range of possible adjuvant systemic toxicity. For instance, it has been suggested that long-term tissue depots formed by aluminum or oil emulsions may have long-term harmful effects. Nevertheless, because disease onset and environmental exposure like immunization can occur decades apart, it can be exceedingly difficult, if not impossible, to detect chronic toxicity and establish any causal relationship.

7.5 Regulatory Approval for Adjuvants and Adjuvanted Vaccines

According to the guidelines provided by the United States Food and Drug Administration (US FDA), it is imperative to ensure that the introduction of an adjuvant into a product is supported by compelling evidence demonstrating its lack of adverse impact on the safety and potency of said product [123]. In accordance with the European Medicines Agency (EMA), it is imperative to substantiate the presence of evidence that substantiates the attainment of enhanced immune response without an unwarranted escalation in both local and systemic adverse reactions [124].

The comprehensive assessment of safety concerns necessitates a profound comprehension of the impact of adjuvants on the immune response and its associated mechanisms. The incorporation of novel scientific data pertaining to vaccine adjuvants necessitates the reassessment of existing regulatory methodologies. The field of adjuvant safety is a crucial and underexplored area of research. Given that adjuvants possess distinct pharmacological properties, it is imperative to consider their potential impact on both the immunogenicity and safety of vaccines. Consequently, a comprehensive evaluation of safety becomes indispensable.

Adjuvants are potential constituents that can be incorporated into vaccine formulations or administered alongside vaccines with the aim of augmenting the immune responses elicited by specific antigen(s) or directing a specific immune response. The adherence to pharmacopoeial requirements is of utmost significance when considering the utilization of adjuvants, as it ensures their compatibility with established standards. Additionally, it is imperative to ascertain that these adjuvants do not induce any form of toxicity that surpasses acceptable thresholds. The manifestation of adjuvant activity is influenced by numerous factors, and it is not generally feasible to extrapolate the immune response achieved with a specific antigen/adjuvant combination to a different antigen [125].

The physical and biological properties of individual antigens exhibit significant variability, thereby leading to distinct characteristics and behaviors. It is worth noting that the interaction between antigens and adjuvants can also manifest in diverse manners, potentially influencing the overall immune response.

Aluminum hydroxide, commonly known as Alum, remains the sole adjuvant authorized by the US FDA for human application. Contrary to the initial statement, it is worth noting that the European Medicines Evaluation Agency has indeed granted licences for three supplementary adjuvants within the last 10 years. According to the EMA, the three adjuvants that have been granted licences are MF59, AS03, and AS03 [126]. One potential strategy to mitigate the constraints imposed by alum is the concurrent administration of alum in conjunction with supplementary adjuvants. AS04 is a formulation that consists of either aluminum hydroxide or aluminum phosphate, combined with the immunostimulatory molecule known as monophosphoryl lipid A (MPL) [127].

Adjuvants, in the context of immunology, are not typically regarded as active constituents and are not authorized for independent application. Instead, they are sanctioned for use solely in conjunction with a particular adjuvant/antigen combination. The rationale behind this approach lies in the fact that the adjuvanted vaccine formulation as a whole is subjected to rigorous evaluation in clinical trials and subsequently utilized for administration post-licensure. The absence of a pre-existing regulatory mandate necessitating phase III trials to establish the added value of an adjuvanted formulation compared to a non-adjuvanted counterpart can be attributed to the fact that certain antigens employed in potential vaccines may not elicit an adequate immune response when administered in isolation. The safety requirements pertaining to the licensure of vaccines necessitate the substantiation of the vaccine's notable absence of deleterious consequences. Furthermore, the assessment of safety must duly account for the inherent attributes of the vaccine in relation to the physiological state of the recipient.

The demonstration of the adjuvant's effect is imperative within the context of preclinical immunogenicity studies. In the event that a novel adjuvant lacks toxicological data, it is imperative to conduct toxicity studies exclusively focused on the adjuvant itself as a preliminary step. The evaluation of novel adjuvants necessitates a comprehensive assessment akin to that conducted for new chemical entities.

The acquisition of these data can be facilitated either from the vaccine manufacturer or by the producer of the adjuvant. To evaluate the safety profile of the adjuvant, it is necessary to conduct an assessment not only of the adjuvant in isolation but also of the potential synergistic adverse effects that may arise from the combination of the antigen and adjuvant in the animal model. When utilizing species-specific proteins such as cytokines as innovative adjuvants, it is imperative to take into account the matter of species-specific response.

The diversity of adjuvants currently in use or under development is attributed to their varied origins and inherent characteristics. In the realm of immunological research, it is noteworthy to mention that aluminum-based adjuvants are composed of elementary inorganic compounds. Additionally, it is worth noting that PLG, which stands for polymeric carbohydrate, serves as a crucial component. Furthermore, it is intriguing to observe that virosomes can be derived from a diverse range of viral particles. Moreover, muramyl dipeptide (MDP; N-acetyl-muramyl-L-alanyl-D-isoglutamine), derived from bacterial cell walls, plays a significant role in this context. It is also pertinent to mention that saponins originate from plants, whereas squalene is derived from the liver of sharks. Lastly, it is worth noting that recombinant endogenous immunomodulators are obtained from recombinant bacterial, yeast, or mammalian cells. Henceforth, it is deemed unsuitable to furnish an exhaustive inventory of specific examinations that ought to be conducted for any given adjuvant or the amalgamation of adjuvant and antigen as per the present set of guidelines.

It is imperative to consider the present guideline in conjunction with the guideline on pharmaceutical and biological aspects of combined vaccines (CPMP/BWP/477/97) [124]. The following instructions should be implemented by manufacturers of adjuvants/vaccines in a manner that is suitable for their specific adjuvant, taking into account the unique circumstances of each case. It is imperative to adhere to the relevant monographs outlined in the European Pharmacopoeia (Ph.Eur.) in order to ensure scientific rigour and adherence to established standards. Committee for Medicinal Products for Human Use and the International Council for Harmonization of Technical Requirements for Pharmaceuticals for Human Use (ICH) have devised stringent guidelines like the guidelines for viral safety (CPMP/ICH/295/95), recombinant DNA (rDNA) proteins (CPMP/ICH/139/95), and cell substrates (CPMP/ICH/294/95) that ensure optimized production of recombinant protein adjuvants and also for utilization of nucleic acid adjuvants. Following these guidelines not only can ensure safe, efficacious, and quality standards for the vaccine adjuvants but also can ease the regulatory burden.

Feasibility and ethical considerations are some of the chief clinical trial–road blockers in evaluating the comparative safety and efficacy of unadjuvanated antigens and their adjuvanated counterparts. Moreover, in the context of enrolling substantial cohorts, it is often the preference of ethical committees or governing bodies to provide the control group with a licenced vaccine that possesses a thoroughly established safety profile. This approach ensures that these individuals derive a benefit from their participation in the study. The establishment of balanced ratios between the vaccine and control groups, specifically adhering to a 1:1 randomization ratio, is highly favored in the assessment of safety outcomes. The inherent challenge in interpreting the data arises from the fact that rare events tend to manifest with greater frequency within a larger group. The evaluation of infrequent untoward occurrences often necessitates the amalgamation or consolidation of findings from multiple well-regulated experiments. Consequently, these experiments must be meticulously devised to warrant the rationalized pooling of data. The utilization of precise statistical methodologies is employed to facilitate the process of pooling. In an ideal scenario, it is recommended to maintain a balanced ratio of vaccine and control recipients in the pooled analysis, preferably at a 1:1 ratio. However, it is crucial to ensure that the exposure to the control product is not less than half of that of the candidate vaccine, at a minimum.

7.5.1 Challenges for Safety Evaluation of Adjuvanated Vaccines

The evaluation of safety in the context of adjuvanted vaccines poses unique challenges. The inclusion of an adjuvant in a formulation may give rise to heightened theoretical apprehension regarding specific negative outcomes, including heightened levels of pro-inflammatory and pyrogenic mediators, heightened local reactogenicity, disruption of self-tolerance, and undesired effects arising from interactions between the antigen and mechanisms induced by the adjuvant.

7.5.2 Challenges During Pre-Clinical Phase

The immunomodulatory properties of adjuvants have been observed to elicit intricate immune responses, although the exact mechanisms underlying their actions remain incompletely understood. The availability of animal models that accurately predict human responses or disease is a rare occurrence. The identification of the mode of action presents significant challenges, not only in terms of achieving accurate results but also in the context of evaluating potential autoimmunity within a suitable model.

7.6 Prospects

The understanding of the principles governing the successful induction of potent and protective immune responses via pain-free delivery systems is improving, and this marks the beginning of an exciting and dynamic period in vaccine research. Findings about the existence of receptors on innate immune cells that identify the distinctive patterns and elements found on pathogens are at the forefront of this research. The identification of pathogen-based ligands for these receptors as prospective next-generation adjuvants to be used in conjunction with recombinant proteins has sparked enthusiasm due to the realization of how crucial innate immune system activation is to the eventual induction of antigen-specific immunity. Much research is being done in this field, which is expected to result in the creation of entirely new classes of vaccine adjuvants with a range of immune response–controlling and manipulating capabilities. These adjuvants might turn out to be strong enough to enable the use of vaccines to treat cancer, autoimmune disorders, and chronic infectious diseases [128, 129]. Nonetheless, the strength of these adjuvants must be carefully weighed against their propensity to overstimulate the immune system, which could have detrimental effects. In this sense, the systemic distribution of adjuvants after administration may be limited by vaccine and adjuvant delivery systems. Furthermore, the primary function of vaccine delivery systems is to guarantee that the adjuvants and antigens interact with the proper immunocompetent cells in an optimal way to elicit the desired responses. Nonetheless, restricting the ability of adjuvants to act on a wide variety of cells may prove to be just as crucial as concentrating their effects on critical immune cells if they are able to trigger strong inflammatory responses. Moreover, nanotechnology platforms and novel delivery techniques have showed clinically promising results and hence are foreseen as licensed future painless adjuvants. However, their costing, storage, and other aspects like toxicities may need further refinement of these formulations before they can be commercially launched. To summarize, it is evident that pharmaceutical scientists have a special and significant role to play in creating the best adjuvants and vaccine delivery systems for systemic and mucosal administration.

7.7 Conclusion

To conclude, adjuvants are essential for improving the effectiveness of vaccines administered non-invasively. Adjuvants provide strong and sustained immune responses without the use of intrusive delivery techniques like injections by promoting innate immune responses and improving antigen presentation, among other mechanisms. This enhances patient adherence and increases immunization program potential, particularly in resource-constrained environments where conventional injection–based methods could provide difficulties. Furthermore, adjuvants' adaptability enables customized immune responses, maximizing vaccine efficacy against a variety of diseases. To completely comprehend the safety profile and long-term implications of adjuvants in non-invasive vaccination delivery, more research is necessary. Notwithstanding these obstacles, the incorporation of adjuvants into non-invasive vaccination approaches has enormous potential to combat infectious illnesses and enhance public health outcomes worldwide.

References

1. Christensen, D., Vaccine adjuvants: Why and how. *Hum. Vaccines Immunother.*, 12, 10, 2709–2711, 2016.
2. Xiang, S.D., Scholzen, A., Minigo, G. *et al.*, Pathogen recognition and development of particulate vaccines: Does size matter? *Methods*, 40, 1, 1–9, 2006.
3. Mitragotri, S., Immunization without needles. *Nat. Rev. Immunol.*, 5, 12, 905–916, 2005.
4. Peek, L.J., Middaugh, C.R., Berkland, C., Nanotechnology in vaccine delivery. *Adv. Drug Deliv. Rev.*, 60, 8, 915–928, 2008.
5. Kane, A., Lloyd, J., Zaffran, M., Simonsen, L., Kane, M., Transmission of hepatitis B, hepatitis C and human immunodeficiency viruses through unsafe injections in the developing world: Model-based regional estimates. *Bull. World Health Organ.*, 77, 10, 801–807, 1999.
6. Yin, Q., Luo, W., Mallajosyula, V. *et al.*, A TLR7-nanoparticle adjuvant promotes a broad immune response against heterologous strains of influenza and SARS-CoV-2. *Nat. Mater.*, 22, 3, 380–390, 2023.
7. Zhao, T., Cai, Y., Jiang, Y. *et al.*, Vaccine adjuvants: mechanisms and platforms. *Signal Transduct. Target. Ther.*, 8, 1, 1–24, 2023.
8. Chen, X., Emerging adjuvants for intradermal vaccination. *Int. J. Pharm.*, 632, August 2022, 122559, 2023.
9. Pulendran B, S., Arunachalam, P., O'Hagan, D.T., Emerging concepts in the science of vaccine adjuvants. *Nat. Rev. Drug Discov.*, 20, 6, 454–475, 2021.
10. Lavelle, E.C. and Ward, R.W., Mucosal vaccines — fortifying the frontiers. *Nat. Rev. Immunol.*, 22, 4, 236–250, 2022.
11. Bielinska, A.U., Janczak, K.W., Landers, J.J. *et al.*, Mucosal immunization with a novel nanoemulsion-based recombinant anthrax protective antigen vaccine protects against Bacillus anthracis spore challenge. *Infect. Immun.*, 75, 8, 4020–4029, 2007.
12. Miquel-Clopés, A., Bentley, E.G., Stewart, J.P., Carding, S.R., Mucosal vaccines and technology. *Clin. Exp. Immunol.*, 196, 2, 205–214, 2019.
13. Nevagi, R.J., Skwarczynski, M., Toth, I., Polymers for subunit vaccine delivery. *Eur. Polym. J.*, 114, 397–410, 2019.
14. Facciolà, A.V., Laganà, G.A., Pietro Di, A., Vaccine adjuvants overview. *Vaccines*, 10, 2022.
15. Parry, N.M., A Review of Transdermal Vaccine Delivery. *Contag. Live*, 2016, [Internet]. Available from: https://www.contagionlive.com/view/a-review-of-transdermal-vaccine-delivery.
16. Gautam, D. and Sahney, R., Reprocessing and reuse of single-use medical devices and the role of interprofessional collaboration: A literature review. *Curr. Med. Res. Pract.*, 10, 2, 70–74, 2020.
17. Skwarczynski, M. and Toth, I., Non-invasive mucosal vaccine delivery: advantages, challenges and the future. *Expert Opin. Drug Deliv.*, 17, 4, 435–437, 2020.

18. Lobaina Mato, Y., Nasal route for vaccine and drug delivery: Features and current opportunities. *Int. J. Pharm.*, 572, 118813, 2019.
19. Kehagia, E., Papakyriakopoulou, P., Valsami, G., Advances in intranasal vaccine delivery: A promising non-invasive route of immunization. *Vaccine*, 41, 24, 3589–3603, 2023.
20. VanBenschoten, H.M. and Woodrow, K.A., Vaginal delivery of vaccines. *Adv. Drug Deliv. Rev.*, 178, 2021.
21. Lu, D. and Hickey, A.J., Pulmonary vaccine delivery. *Expert Rev. Vaccines*, 6, 2, 213–226, 2007.
22. Miquel-Clopés, A., Bentley, E.G., Stewart, J.P., Carding, S.R., Mucosal vaccines and technology. *Clin. Exp. Immunol.*, 196, 2, 205–214, 2019.
23. Mangla, B., Javed, S., Sultan, M.H., Ahsan, W., Aggarwal, G., Kohli, K., Nanocarriers-Assisted Needle-Free Vaccine Delivery Through Oral and Intranasal Transmucosal Routes: A Novel Therapeutic Conduit. *Front. Pharmacol.*, 12, 1–20, 2022.
24. Zhu, Q., Chen, Z., Paul, P.K., Lu, Y., Wu, W., Qi, J., Oral delivery of proteins and peptides: Challenges, status quo and future perspectives. *Acta Pharm. Sin. B*, 11, 8, 2416–2448, 2021.
25. Jia, W., Zhang, T., Huang, H. *et al.*, Colorectal cancer vaccines: The current scenario and future prospects. *Front. Immunol.*, 13, 942235, 2022.
26. Kozlowski, P.A., Cu-Uvin, S., Neutra, M.R., Flanigan, T.P., Comparison of the oral, rectal, and vaginal immunization routes for induction of antibodies in rectal and genital tract secretions of women. *Infect. Immun.*, 65, 4, 1387–1394, 1997.
27. Facciolà, A., Visalli, G., Laganà, P. *et al.*, The new era of vaccines: The "nanovaccinology." *Eur. Rev. Med. Pharmacol. Sci.*, 23, 16, 7163–7182, 2019.
28. Olivieri, B., Betterle, C., Zanoni, G., Vaccinations and autoimmune diseases. *Vaccines*, 9, 8, 1–15, 2021.
29. Kumar, A. and Kumar, A., Mucosal and transdermal vaccine delivery strategies against COVID-19. *Drug Deliv. Transl. Res.*, 12, 5, 968–972, 2022.
30. Ita, K., Transdermal delivery of vaccines – Recent progress and critical issues. *Biomed. Pharmacother.*, 83, 1080–1088, 2016.
31. Ho, A.W. and Kupper, T.S., T cells and the skin: from protective immunity to inflammatory skin disorders. *Nat. Rev. Immunol.*, 19, 8, 490–502, 2019.
32. Newton, K. and Dixit, V.M., Signaling in innate immunity and inflammation. *Cold Spring Harb. Perspect. Biol.*, 4, 3, 1–19, 2012.
33. Chavda, V.P., Teli, D., Balar, P.C. *et al.*, Self-assembled peptide hydrogels for the treatment of diabetes and associated complications. *Colloids Surf. B Biointerfaces*, 235, 113761, 2024.
34. Ahmed, S., Amin, M.M., Sayed, S., Ocular Drug Delivery: a Comprehensive Review. *AAPS PharmSciTech.*, 24, 2, 66, 2023.
35. Krishnaswami, V., Kandasamy, R., Alagarsamy, S., Palanisamy, R., Natesan, S., Biological macromolecules for ophthalmic drug delivery to treat ocular diseases. *Int. J. Biol. Macromol.*, 110, 7–16, 2018.

36. Tannock, G.A., Paul, J.A., Barry, R.D., Immunization against influenza by the ocular route. *Vaccine*, 3, 3, 277–280, 1985.
37. Seo, K.Y., Han, S.J., Cha, H.-R. *et al.*, Eye Mucosa: An Efficient Vaccine Delivery Route for Inducing Protective Immunity. *J. Immunol.*, 185, 6, 3610–3619, 2010.
38. Lee, J.S., Yoon, S., Han, S.J. *et al.*, Eyedrop vaccination: an immunization route with promises for effective responses to pandemics. *Expert Rev. Vaccines*, 21, 1, 91–101, 2021, https://doi.org/10.1080/14760584.2022.2008246.
39. Facciolà, A., Visalli, G., Laganà, A., Di Pietro, A., An Overview of Vaccine Adjuvants: Current Evidence and Future Perspectives. *Vaccines*, 10, 5, 1–26, 2022.
40. Baylor, N.W., Egan, W., Richman, P., Aluminum salts in vaccines - US perspective. *Vaccine*, 20, SUPPL. 3, 18–23, 2002.
41. He, P., Zou, Y., Hu, Z., Advances in aluminum hydroxide-based adjuvant research and its mechanism. *Hum. Vaccines Immunother.*, 11, 2, 477–488, 2015.
42. Li, M., Wang, Y., Sun, Y., Cui, H., Zhu, S.J., Qiu, H.J., Mucosal vaccines: Strategies and challenges. *Immunol. Lett.*, 217, 116–125, 2020.
43. Clapp, T., Siebert, P., Chen, D., Jones Braun, L.T., Vaccines with aluminum-containing adjuvants: Optimizing vaccine efficacy and thermal stability. *J. Pharm. Sci.*, 100, 2, 388–401, 2011.
44. Hem, S.L. and Hogenesch, H., Relationship between physical and chemical properties of aluminum-containing adjuvants and immunopotentiation. *Expert Rev. Vaccines*, 6, 5, 685–698, 2007.
45. Lawson, L.B., Norton, E.B., Clements, J.D., Defending the mucosa: Adjuvant and carrier formulations for mucosal immunity. *Curr. Opin. Immunol.*, 23, 3, 414–420, 2011.
46. Lycke, N. and Lebrero-Fernández, C., ADP-ribosylating enterotoxins as vaccine adjuvants. *Curr. Opin. Pharmacol.*, 41, 42–51, 2018.
47. Livares, J.C.A. and Ood, J.W., Review article West Nile virus infection of horses. *Vet. Res.*, 35, 467–483, 2004.
48. Kayamuro, H., Yoshioka, Y., Abe, Y. *et al.*, Interleukin-1 Family Cytokines as Mucosal Vaccine Adjuvants for Induction of Protective Immunity against Influenza Virus. *J. Virol.*, 84, 24, 12703–12712, 2010.
49. Mosaheb, M.M., Reiser, M.L., Wetzler, L.M., Toll-like receptor ligand-based vaccine adjuvants require intact MyD88 signaling in antigen-presenting cells for germinal center formation and antibody production. *Front. Immunol.*, 8, 1–12, MAR, 2017.
50. USFDA. https://www.fda.gov/vaccines-blood-biologics/vaccines/biothrax [Internet], 2023. Available from: https://www.fda.gov/vaccines-blood-biologics/vaccines/biothrax.
51. USFDA. https://www.fda.gov/vaccines-blood-biologics/vaccines/cervarix [Internet], 2019. Available from: https://www.fda.gov/vaccines-blood-biologics/vaccines/cervarix.

52. EMA. https://www.ema.europa.eu/en/medicines/human/EPAR/ixiaro [Internet], 2019. Available from: https://www.ema.europa.eu/en/medicines/human/EPAR/ixiaro.
53. Jones, D.H., Menjugate (Chiron). *Curr. Opin. Investig. Drugs (Lond. Engl. 2000)*, 2, 1, 47–49, 2001.
54. Song, K.R., Lim, J.K., Park, S.E. *et al.*, Oral cholera vaccine efficacy and effectiveness. *Vaccines*, 9, 12, 1–18, 2021.
55. USFDA. https://www.fda.gov/vaccines-blood-biologics/fluad-quadrivalent [Internet], 2021. Available from: https://www.fda.gov/vaccines-blood-biologics/fluad-quadrivalent.
56. Li, M., Wang, Y., Sun, Y., Cui, H., Zhu, S.J., Qiu, H.J., Mucosal vaccines: Strategies and challenges. *Immunol. Lett.*, 217, 116–125, 2020.
57. Peek, L.J., Middaugh, C.R., Berkland, C., Nanotechnology in vaccine delivery. *Adv. Drug Deliv. Rev.*, 60, 8, 915–928, 2008.
58. Zhao, L., Seth, A., Wibowo, N. *et al.*, Nanoparticle vaccines. *Vaccine*, 32, 3, 327–337, 2014.
59. Chavda, V.P., Nalla, L.V., Balar, P. *et al.*, Advanced Phytochemical-Based Nanocarrier Systems for the Treatment of Breast Cancer. *Cancers*, 15, 4, 1023, 2023.
60. Bonam, S.R., Partidos, C.D., Halmuthur, S.K.M., Muller, S., An Overview of Novel Adjuvants Designed for Improving Vaccine Efficacy. *Trends Pharmacol. Sci.*, 38, 9, 771–793, 2017.
61. Agger, E.M., Novel adjuvant formulations for delivery of anti-tuberculosis vaccine candidates. *Adv. Drug Deliv. Rev.*, 102, 73–82, 2016.
62. Schmidt, S.T., Foged, C., Korsholm, K.S., Rades, T., Christensen, D., Liposome-based adjuvants for subunit vaccines: Formulation strategies for subunit antigens and immunostimulators. *Pharmaceutics*, 8, 1, 1–22, 2016.
63. Márquez-Escobar, V.A., Bañuelos-Hernández, B., Rosales-Mendoza, S., Expression of a Zika virus antigen in microalgae: Towards mucosal vaccine development. *J. Biotechnol.*, 282, 86–91, 2018.
64. Schwendener, R.A., Liposomes as vaccine delivery systems: A review of the recent advances. *Ther. Adv. Vaccines*, 2, 6, 159–182, 2014.
65. McKee, A.S. and Marrack, P., Old and new adjuvants. *Curr. Opin. Immunol.*, 47, 44–51, 2017.
66. Data, R.U.S.A. and Kishore, P.E.S., United States Patent. *Patent.* 1, 12, 2002.
67. Dai, C.C., Yang, J., Hussein, W.M. *et al.*, Polyethylenimine: An Intranasal Adjuvant for Liposomal Peptide-Based Subunit Vaccine against Group A Streptococcus. *ACS Infect. Dis.*, 6, 9, 2502–2512, 2020.
68. Varypataki, E.M., van der Maaden, K., Bouwstra, J., Ossendorp, F., Jiskoot, W., Cationic Liposomes Loaded with a Synthetic Long Peptide and Poly(I:C): a Defined Adjuvanted Vaccine for Induction of Antigen-Specific T Cell Cytotoxicity. *AAPS J.*, 17, 1, 216–226, 2015.
69. Witika, B.A., Bassey, K.E., Demana, P.H., Siwe-Noundou, X., Poka, M.S., Current Advances in Specialised Niosomal Drug Delivery: Manufacture,

Characterization and Drug Delivery Applications. *Int. J. Mol. Sci.*, 23, 17, 1–35, 2022.
70. Rajput, T. and Chauhan, M.K., Bilosome: a Bile Salt Based Novel Carrier System Gaining Interest in Pharmaceutical Research. *J. Drug Deliv. Ther.*, 7, 5, 4–16, 2017.
71. Donaldson, B., Lateef, Z., Walker, G.F., Young, S.L., Ward, V.K., Virus-like particle vaccines: immunology and formulation for clinical translation. *Expert Rev. Vaccines*, 17, 9, 833–849, 2018.
72. Salvador, A., Igartua, M., Hernández, R.M., Pedraz, J.L., View this Special Issue Academic Editor: Sanyog Jain. *J. Drug. Deliv.*, 2011, 1–40, 2011.
73. Plummer, E.M. and Manchester, M., Viral nanoparticles and virus-like particles: Platforms for contemporary vaccine design. *Wiley Interdiscip. Rev. Nanomed. Nanobiotechnol.*, 3, 2, 174–196, 2011.
74. Moser, C., Amacker, M., Zurbriggen, R., Influenza virosomes as a vaccine adjuvant and carrier system. *Expert Rev. Vaccines*, 10, 4, 437–446, 2011.
75. Shafique, M., Meijerhof, T., Wilschut, J., de Haan, A., Evaluation of an Intranasal Virosomal Vaccine against Respiratory Syncytial Virus in Mice: Effect of TLR2 and NOD2 Ligands on Induction of Systemic and Mucosal Immune Responses. *PLoS One*, 8, 4, 1–19, 2013.
76. Specht, E.A. and Mayfield, S.P., Algae-based oral recombinant vaccines. *Front. Microbiol.*, 5, FEB, 1–7, 2014.
77. Bezerra, I.P.S., Costa-Souza, B.L.S., Carneiro, G., Ferreira, L.A.M., de Matos Guedes, H.L., Rossi-Bergmann, B., Nanoencapsulated retinoic acid as a safe tolerogenic adjuvant for intranasal vaccination against cutaneous leishmaniasis. *Vaccine*, 37, 28, 3660–3667, 2019.
78. Erasmus, J.H., Khandhar, A.P., Guderian, J. *et al.*, A Nanostructured Lipid Carrier for Delivery of a Replicating Viral RNA Provides Single, Low-Dose Protection against Zika. *Mol. Ther.*, 26, 10, 2507–2522, 2018.
79. Shah, R.R., Dodd, S., Schaefer, M. *et al.*, The Development of Self-Emulsifying Oil-in-Water Emulsion Adjuvant and an Evaluation of the Impact of Droplet Size on Performance. *J. Pharm. Sci.*, 104, 4, 1352–1361, 2015.
80. Zhang, J., Miao, J., Han, X. *et al.*, Development of a novel oil-in-water emulsion and evaluation of its potential adjuvant function in a swine influenza vaccine in mice. *BMC Vet. Res.*, 14, 1, 1–18, 2018.
81. Chen, Y.S., Hung, Y.C., Lin, W.H., Huang, G.S., Assessment of gold nanoparticles as a size-dependent vaccine carrier for enhancing the antibody response against synthetic foot-and-mouth disease virus peptide. *Nanotechnology*, 21, 19, 1–9, 2010.
82. Zhao, W., Wu, W., Xu, X., Oral vaccination with liposome-encapsulated recombinant fusion peptide of urease B epitope and cholera toxin B subunit affords prophylactic and therapeutic effects against H. pylori infection in BALB/c mice. *Vaccine*, 25, 44, 7664–7673, 2007.
83. Khan, O.F., Cooper, C.L., Mcpartlan, J.S. *et al.*, Erratum: Dendrimer-RNA nanoparticles generate protective immunity against lethal Ebola, H1N1

influenza, and Toxoplasma gondii challenges with a single dose (Proceedings of the National Academy of Sciences of the United States of America (2016) 113 (E4133-. *Proc. Natl. Acad. Sci. U.S.A.*, 113, 35, E5250, 2016.
84. Bhatia, S., Natural Polymers vs Synthetic Polymer, in: *Natural Polymer Drug Delivery Systems*, pp. 95–118, 2016.
85. Roopngam, P.E., Liposome and polymer-based nanomaterials for vaccine applications. *Nanomedicine J.*, 6, 1–10, 2019.
86. Nevagi, R.J., Skwarczynski, M., Toth, I., Polymers for subunit vaccine delivery. *Eur. Polym. J.*, 114, 397–410, 2019.
87. Foged, C., Hansen, J., Agger, E.M., License to kill: Formulation requirements for optimal priming of CD8 + CTL responses with particulate vaccine delivery systems. *Eur. J. Pharm. Sci.*, 45, 4, 482–491, 2012.
88. Singh, B., Maharjan, S., Sindurakar, P., Cho, K.H., Choi, Y.J., Cho, C.S., Needle-free immunization with chitosan-based systems. *Int. J. Mol. Sci.*, 19, 11, 1–27, 2018.
89. Diego-González, L., Crecente-Campo, J., Paul, M.J. *et al.*, Design of polymeric nanocapsules for intranasal vaccination against mycobacterium tuberculosis: Influence of the polymeric shell and antigen positioning. *Pharmaceutics*, 12, 6, 1–25, 2020.
90. Nevagi, R.J., Khalil, Z.G., Hussein, W.M. *et al.*, Polyglutamic acid-trimethyl chitosan-based intranasal peptide nano-vaccine induces potent immune responses against group A streptococcus. *Acta Biomater.*, 80, 278–287, 2018.
91. Dhakal, S., Renu, S., Ghimire, S. *et al.*, Mucosal immunity and protective efficacy of intranasal inactivated influenza vaccine is improved by chitosan nanoparticle delivery in pigs. *Front. Immunol.*, 9, 1–16, MAY, 2018.
92. Bi, C.C., Wang, A.P., Chu, Y.C. *et al.*, Intranasal delivery of rotigotine to the brain with lactoferrin-modified PEG-PLGA nanoparticles for Parkinson's disease treatment. *Int. J. Nanomed.*, 11, 6547–6559, 2016.
93. Ebrahimian, M., Hashemi, M., Maleki, M. *et al.*, Induction of a balanced Th1/Th2 immune responses by co-delivery of PLGA/ovalbumin nanospheres and CpG ODNs/PEI-SWCNT nanoparticles as TLR9 agonist in BALB/c mice. *Int. J. Pharm.*, 515, 1–2, 708–720, 2016.
94. Dewangan, H.K., Pandey, T., Maurya, L., Singh, S., Rational design and evaluation of HBsAg polymeric nanoparticles as antigen delivery carriers. *Int. J. Biol. Macromol.*, 111, 804–812, 2018.
95. Tamayo, I., Irache, J.M., Mansilla, C., Ochoa-Repáraz, J., Lasarte, J.J., Gamazo, C., Poly(anhydride) nanoparticles act as active Th1 adjuvants through toll-like receptor exploitation. *Clin. Vaccine Immunol.*, 17, 9, 1356–1362, 2010.
96. Grego, E., Si, A., Metin, U. *et al.*, Polymeric Nanoparticle-Based Vaccine Adjuvants and Delivery Vehicles. *Curr. Top Microbiol. Immunol.*, 433, 29–76, 2021.
97. Zaver, S.A. and Woodward, J.J., Cyclic dinucleotides at the forefront of innate immunity. *Curr. Opin. Cell Biol.*, 63, 49–56, 2020.

98. Chaung, H.-C., CpG oligodeoxynucleotides as DNA adjuvants in vertebrates and their applications in immunotherapy. *Int. Immunopharmacol.*, 6, 10, 1586–1596, 2006.
99. Krishnan, L., Dicaire, C.J., Patel, G.B., Sprott, G.D., Archaeosome vaccine adjuvants induce strong humoral, cell-mediated, and memory responses: Comparison to conventional liposomes and alum. *Infect. Immun.*, 68, 1, 54–63, 2000.
100. Burt, D., Mallett, C., Plante, M., Zimmermann, J., Torossian, K., Fries, L., Proteosome-adjuvanted intranasal influenza vaccines: Advantages, progress and future considerations. *Expert Rev. Vaccines*, 10, 3, 365–375, 2011.
101. Lambkin-Williams, R., Gelder, C., Broughton, R. *et al.*, An intranasal Proteosome-adjuvanted trivalent influenza vaccine is safe, immunogenic & efficacious in the human viral influenza challenge model. Serum IgG & mucosal IgA are important correlates of protection against illness associated with infection. *PLoS One*, 11, 12, e0163089, 2016.
102. Sarasati, A., Wihadmadyatami, H., Ana, I.D., Carbonate apatite nanoparticles: A novel nano-adjuvant for oral mucosal vaccines and immunomodulator. *OpenNano*, 12, April, 100149, 2023.
103. Kim, K.S., Kim, H., Park, Y. *et al.*, Noninvasive Transdermal Vaccination Using Hyaluronan Nanocarriers and Laser Adjuvant. *Adv. Funct. Mater.*, 26, 15, 2512–2522, 2016.
104. Chavda, V., Zajac, K.K., Gunn, J.L. *et al.*, Ethnic differences in hepatocellular carcinoma prevalence and therapeutic outcomes. *Cancer Rep.*, 6(Suppl 1), e1821, 2023.
105. Hu, K., Dou, J., Yu, F. *et al.*, An ocular mucosal administration of nanoparticles containing DNA vaccine pRSC-gD-IL-21 confers protection against mucosal challenge with herpes simplex virus type 1 in mice. *Vaccine*, 29, 7, 1455–1462, 2011.
106. Rossi, I., Spagnoli, G., Buttini, F. *et al.*, A respirable HPV-L2 dry-powder vaccine with GLA as amphiphilic lubricant and immune-adjuvant. *J.Control. Release Off. J. Control. Release Soc.*, 340, 209–220, 2021.
107. McNeilly, C.L., Crichton, M.L., Primiero, C.A., Frazer, I.H., Roberts, M.S., Kendall, M.A.F., Microprojection arrays to immunise at mucosal surfaces. *J. Control. Release*, 196, 252–260, 2014.
108. White, J.A., Blum, J.S., Hosken, N.A. *et al.*, Serum and mucosal antibody responses to inactivated polio vaccine after sublingual immunization using a thermoresponsive gel delivery system. *Hum. Vaccines Immunother.*, 10, 12, 3611–3621, 2014.
109. Chavda, V.P., Balar, P.C., Teli, D., Davidson, M., Bojarska, J., Apostolopoulos, V., Antibody–Biopolymer Conjugates in Oncology: A Review. *Molecules*, 28, 6, 2605, 2023.
110. Kelly, S.H., Wu, Y., Varadhan, A.K., Curvino, E.J., Chong, A.S., Collier, J.H., Enabling sublingual peptide immunization with molecular self-assemblies. *Biomaterials*, 241, January, 119903, 2020.

111. Seth, A., Kong, I.G., Lee, S.H. *et al.*, Modular virus-like particles for sublingual vaccination against group A streptococcus. *Vaccine*, 34, 51, 6472–6480, 2016.
112. Mattila, J.P., Mirandola, L., Chiriva-Internati, M., Development of a M cell-targeted microparticulate platform, BSK02™, for oral immunization against the ovarian cancer antigen, sperm protein 17. *J. Biomed. Mater. Res. - Part B Appl. Biomater.*, 107, 1, 29–36, 2019.
113. Zhang, Y., Li, H., Wang, Q. *et al.*, Rationally Designed Self-Assembling Nanoparticles to Overcome Mucus and Epithelium Transport Barriers for Oral Vaccines against Helicobacter pylori. *Adv. Funct. Mater.*, 28, 33, 1–15, 2018.
114. Moon, J.J., Suh, H., Bershteyn, A. *et al.*, Interbilayer-crosslinked multilamellar vesicles as synthetic vaccines for potent humoral and cellular immune responses. *Nat. Mater.*, 10, 3, 243–251, 2011.
115. Si, Y., Wen, Y., Kelly, S.H., Chong, A.S., Collier, J.H., Intranasal delivery of adjuvant-free peptide nanofibers elicits resident CD8+ T cell responses. *J. Control. Release*, 282, 120–130, 2018.
116. Ho, H.M., Huang, C.Y., Cheng, Y.J. *et al.*, Squalene nanoemulsion reinforces mucosal and immunological fingerprints following intravaginal delivery. *Biomed. Pharmacother.*, 141, June, 111799, 2021.
117. Suk, J.S., Xu, Q., Kim, N., Hanes, J., Ensign, L.M., PEGylation as a strategy for improving nanoparticle-based drug and gene delivery. *Adv. Drug Deliv. Rev.*, 99, 28–51, 2016.
118. Petrovsky, N., Freeing vaccine adjuvants from dangerous immunological dogma. *Expert Rev. Vaccines*, 7, 1, 7–10, 2008.
119. Petrovsky, N. and Aguilar, J.C., Vaccine adjuvants: Current state and future trends. *Immunol. Cell Biol.*, 82, 5, 488–496, 2004.
120. Leventhal, J.S., Berger, E.M., Brauer, J.A., Cohen, D.E., Hypersensitivity Reactions to Vaccine Constituents: A Case Series and Review of the Literature. *Dermatitis*, 23, 3, 102–109, 2012.
121. Powell, B.S., Andrianov, A.K., Fusco, P.C., Polyionic vaccine adjuvants: another look at aluminum salts and polyelectrolytes. *Clin. Exp. Vaccine Res.*, 4, 1, 23, 2015.
122. Chérin, P., Laforet, P., Ghérardi, R.K. *et al.*, La myofasciite à macrophages: description, hypothèses étiopathogéniques. *Revue de Médecine Interne*, 20, 6, 483–489, 1999.
123. Ripley, S., Guidance for industry for the evaluation of combination vaccines for preventable diseases: production, testing and clinical studies, FDA-1997-N-0454.
124. EMA. Committee for Medicinal Products for Human Use (Chmp), Guideline on Adjuvants in Vaccines for Human Use Guideline on Adjuvants in Vaccines. *Agency*, 7, December 2004, E35, 2005.

125. *Draft Regulatory Guidelines For Development of Vaccines With Special Consideration Covid-19 Vaccine*, Central Drugs Standard Control Organization Directorate General of Health Services Ministry of Health and Family Welfare Government of India, Kotla Road, New Delhi, pp. 1–40, 2020.
126. Hassett, K.J., Cousins, M.C., Rabia, L.A. *et al.*, Stabilization of a recombinant ricin toxin A subunit vaccine through lyophilization. *Eur. J. Pharm. Biopharm.*, 85, 2, 279–286, 2013.
127. Garçon, N., Chomez, P., Van Mechelen, M., GlaxoSmithKline Adjuvant Systems in vaccines: Concepts, achievements and perspectives. *Expert Rev. Vaccines*, 6, 5, 723–739, 2007.
128. Chavda, V.P., Sugandhi, V.V., Pardeshi, C.V. *et al.*, Current status of Cancer Nanotheranostics: Emerging strategies for cancer management. *J. Drug Deliv. Sci. Technol.*, 85, 104579, 2023.
129. Chavda, V.P., Sugandhi, V.V., Pardeshi, C.V. *et al.*, Engineered exosomes for cancer theranostics: Next-generation tumor targeting. *J. Drug Deliv. Sci. Technol.*, 85, 104579, 2023.

8

Regulatory Guidelines for Vaccine Adjuvants

Suneetha Vuppu[1]*, Vivek P. Chavda[2†], Toshika Mishra[1], Nikita Sharma[1] and Sathvika Kamaraj[1]

[1]Department of Biotechnology, School of Bio Sciences and Technology, Vellore Institute of Technology, Vellore, Tamil Nadu, India
[2]Department of Pharmaceutics and Pharmaceutical Technology, L.M. College of Pharmacy, Ahmedabad, Gujarat, India

Abstract

Vaccine adjuvants are a diverse set of naturally occurring or produced chemicals that enhance the immunological response to the antigen. Adjuvants may be used into vaccination candidates to improve the efficiency of weak antigens and generate appropriate immune responses. A number of factors can influence adjuvant and formulation selection, including the chemical and physical characteristics of the vaccine antigen, the type of immune response desired, the age of the target population, and the manner of vaccine administration. Adjuvants such as alum, virosomes, MPL, AS01, AS04, and RC-529 are approved and employed in production of various types of vaccines. Other platforms of adjuvants also include saponin-based, GalCer-based, lipopolysaccharide-based, zwitterionic polysaccharide–based, and Toll-like receptor agonist molecule–based adjuvants. The adjuvants can function as immunostimulant or delivery system mediating controlled release of antigen for a prolonged period of time. The mechanism of adjuvanticity involves depot formation for protection of antigen, enhanced presentation of antigen, and modulation of immune response. The subject of vaccine development is constantly evolving due to ongoing research and development of novel vaccines in numerous infectious illnesses, including those causing infectious diarrhea. An adjuvant is used to enhance the immunological reaction to an antigen in a vaccine. Given the regulatory and public scrutiny that surrounds the use of adjuvants, it

**Corresponding author*: vsuneetha@vit.ac.in; ORCID: https://orcid.org/0000-0002-4694-1944
†*Corresponding author*: Vivek7chavda@gmail.com; ORCID: https://orcid.org/0000-0002-7701-8597

Vivek P. Chavda and Vasso Apostolopoulos (eds.) Emerging Pathways of Vaccine Adjuvants: A Nonspecific Stimulant of the Immune System, (211–234) © 2025 Scrivener Publishing LLC

is imperative to present a data-driven rationale for their usage in an appropriate animal model. International authorities such as the World Health Organisation, Food and Drug Administration, European Medicines Agency, Health Canada, and Australian Therapeutic Goods Administration establish regulatory rules for this. Adjuvants must be evaluated and regulated for immunogenicity and toxicity because not all adjuvants are designed to cause the same type or duration of immune system activation. Thus, regulatory agencies and health authorities issue regulatory recommendations to guarantee the safety and efficiency of adjuvant-containing vaccinations.

Keywords: Vaccines, adjuvants, immunostimulant, carrier, antigen-adjuvant, regulation, guidelines

8.1 Introduction

The development of vaccines is considered to be one of the greatest contributions to medicine [1]. Vaccines are considered as an advanced biological mechanisms crafted to stimulate the human immune system against harmful pathogens [2]. These biological formulations promote immunity without actually triggering the disease by helping the immune system identify and fight against certain invaders. Vaccine development exemplifies the constant pursuit of scientific innovations. Vaccines have prevented countless deaths worldwide by helping to manage and eradicate fatal illnesses throughout time. The process of developing them is intricate and usually begins with a thorough investigation of the characteristics of the pathogenic organism that is being targeted. To assure safety and efficacy, the complex process of developing vaccines includes extensive research, preclinical studies, clinical trials, and regulatory reviews. As the globe faces new health concerns, continued vaccination research and development activities are vital for safeguarding people from both known and unknown infectious agents. Subsequent preclinical investigations using animal models and lab settings offer essential information on immunogenicity and safety [3, 4].

Vaccines are then tested in three stages of clinical trials in humans to determine their safety, dosage, and effectiveness. Regulatory bodies meticulously examine data prior to granting approval. Remarkably, recent years have seen innovations, such as mRNA and vector-based platforms, bringing about a revolution in the field of vaccine development [5]. Collaborative efforts have intensified due to the urgency imposed by growing infectious threats like as the COVID-19 pandemic, demonstrating the flexibility and resilience of vaccine research and development [6–8]. The process of developing vaccines is complex and dynamic, including scientific advancement, stringent testing, and regulatory oversight. The initial stages of vaccine development involve comprehensive research to understand the biological nuances of the targeted pathogen, identify potential antigens, and devise a formulation capable of inducing a robust immune response [9]. This phase integrates molecular biology, virology, and immunology to unravel the intricacies of the infectious agents. Preclinical investigations are then conducted, which include laboratory research and animal testing to determine the vaccine candidate's safety and efficacy. Before proceeding to human trials, this step assists researchers in identifying potential challenges and refining the formulation. Clinical trials move through distinct phases. Phase I centers on safety, phase II examines efficacy and optimal dosage, and phase III includes large-scale trials for effectiveness confirmation and monitoring rare side effects [10]. Regulatory bodies, like the U.S. Food and Drug Administration (FDA) and the European Medicines Agency (EMA), rely on the data from these trials to make informed decisions regarding approval [11]. Innovative technologies such as mRNA and viral vector platforms have evolved in recent years, enabling more rapid development and manufacturing. Particularly during the COVID-19 pandemic, these technologies were quickly deployed to produce vaccinations efficiently. The efficacy of vaccine development is intertwined with the collaborative efforts of researchers, pharmaceutical companies, regulatory bodies, and global health organizations [12]. The efficacy of vaccine development is intertwined with the collaborative efforts of researchers, pharmaceutical companies, regulatory bodies, and global health organizations. The development of vaccines poses a significant ongoing challenge, especially in regions characterized by high endemicity or among populations at an increased risk, seeking to thwart the spread of infections [13].

Vaccine development offers several advantages in disease prevention along with health promotion. The use of vaccines contributes significantly to the prevention of infectious diseases, which lowers the incidence of serious illnesses and improves health globally [14]. They help to build herd immunity, which protects not just individuals but also susceptible communities [15]. Vaccination campaigns have been shown to be cost-effective, resulting in considerable savings in healthcare expenses. The effectiveness of vaccinations has proven vital in maintaining the security of the global health system by helping to control the spread of newly identified contagious illnesses [1]. In addition to potentially lowering antibiotic resistance, effective vaccination campaigns will help in eradicate several illnesses. The multifaceted benefits of vaccine development go beyond individual protection, offering broader advantages for health, community resilience, and the overall well-being of the global population.

8.2 Vaccine Adjuvants

Vaccination has proven to be highly effective in preventing numerous infectious diseases, and preserving millions of lives. Vaccines have been essential in mitigating the impact of infectious diseases. Over the years, we have witnessed multiple developments in isolating organisms, cell culture, genomic sequencing, and recombinant and nuclear technologies. These strategies have significantly enhanced the advancement of vaccines, resulting in immunization [16]. The fundamental premise of vaccination is to stimulate an immune response that can offer targeted protection against infectious diseases while minimizing or eliminating the risk of contracting the disease through vaccination. However, there is a requirement to introduce an additional natural immune response in conjunction with the vaccine. Consequently, adjuvants were added to the vaccine antigen in order to introduce immunostimulatory components and increase the immunogenicity of the immunogen.

Adjuvants are essential elements in vaccines that are now considered ubiquitous. Vaccine adjuvants are a heterogeneous group of many naturally occurring or synthesized substances that boost immune response against the antigen [17]. Developers may incorporate adjuvants into vaccination candidates to augment the effectiveness of feeble antigens and elicit adequate immune responses. The choice of adjuvant and formulation might be dependent on various parameters, such as the physical and chemical characteristics of the vaccine antigen, desired immune response type, target population age, and vaccine delivery route [18]. Various adjuvants have been approved by regulatory bodies employed, including alum, virosomes, MPL, AS01, AS04, and RC-529 [19].

Aluminum-containing adjuvants are the most prevalent metals occurring naturally and can be found in the environment. Adjuvants containing alum have been incorporated into vaccines to defend the body against pathogens. Monophosphoryl lipid A is a non-alum vaccination adjuvant with global clinical and market approval. The efficacy of adjuvant, monophosphoryl lipid A has been primarily attributed to its activity as an agonist of Toll-like receptor 4 (TLR4), that is responsible for the direct stimulation of dendritic cells. This is because monophosphoryl lipid A is an exceedingly purified type of lipopolysaccharide (LPS) present in the cell wall of *Salmonella enterica.*

AS01 is a vaccination adjuvant system that consists of liposomes and contains two immunostimulants: 3-O-desacyl-4′-monophosphoryl lipid A and saponin QS-21. The impact of AS01 is immediate and short-term, limited to the muscle where it was injected and the nearby lymph node where it discharges. AS01 is highly effective in stimulating CD4+ T-cell–immunity responses and is a suitable adjuvant alternative to vaccines designed to target viruses or intracellular pathogens [20]. AS04 comprises 3-O-desacyl-4′-monophosphoryl lipid A derived from *Salmonella minnesota*. This modified form of LPS is attached to aluminum salts through adsorption. AS04's adjuvant effect is achieved via activating TLR4 on innate cells, in combination with the immunomodulatory actions of alum [19]. RC-529 acts as a potent vaccine adjuvant. The selective phosphorylation, glycosylation, and acylation processes can be carried out with the synthetic strategy's adaptable application of orthogonal protecting groups [21].

The development of adjuvant formulations primarily centers around platforms composed of micro- and nanoparticles, such as emulsions, liposomes, and aluminum compounds [18]. Varying formulations of identical immunomodulatory molecules have the potential to elicit significantly distinct immune responses. Six of the seven vaccine recipients were successfully vaccinated against malaria using repeat T-cell epitope surface antigen (RTS), S vaccine potential candidate, which was produced with AS02. This phenomenon was demonstrated in the malaria vaccine program. Conversely, the identical antigen protected only two out of the 7 recipients when administered with AS03/4, respectively [22].

It was determined that an H5N1 pandemic vaccine with a viable adjuvant system could be developed with adequate immunogenicity to facilitate antigen-sparing while preserving cross-reactivity in all age groups. The efficacy of the AS03 Adjuvant System in surmounting significant clinical obstacles in developing a pandemic vaccine was demonstrated by formulating a candidate H5N1 strain [23]. A study by Einstein *et al.* (2009) compared the immunological response elicited by the licensed aluminum salt-adjuvanted HPV6/11/16/18 vaccine and the AS04-adjuvanted HPV16/18 vaccine in women aged 18–45. The results indicated that the HPV-16/18-AS04 vaccine produced more significant quantities of memory B cells and neutralizing antibodies against HPV-16 and HPV-18 [24]. Moreover, immunization with HSV/AS04 provided sufficient defense against primary HSV illness and was associated with a reduction in the spread of the virus in the genital tract. However, it did not successfully prevent mucosal infection [25].

8.3 Mechanism of Action

Currently, our understanding of vaccination adjuvant mechanisms is centered on innate immunity mediating antigen-specific responses. The adjuvants can be utilized in vaccine formulation as immunostimulants or delivery systems. As an immunostimulant, the adjuvants interact with TLRs as well as other pattern recognition receptors (PRRs) causing maturation and activation of antigen-presenting cells (APCs). The adjuvants as delivery system modulate presentation of antigen extending the bioavailability of antigens loaded and facilitating their transport to APCs or lymph nodes [17].

Studies have reported that the mechanism of adjuvanticity involves three important steps.

8.3.1 Formation of Depot for Antigen Protection

The carrier adjuvant molecules containing entrapped, adsorbed, or aggregated antigens facilitate slow release of antigen for a prolonged time period. This is referred to as the "depot effect", which enhances vaccination immunogenicity by preventing the liver from clearing antigen. Particulate carrier molecules such as virosome, liposome, emulsion, cochleates, lipid, as well as polymeric nanoparticles functions as carrier or protector delivering the vaccine. This contributes to the immunostimulatory function of the adjuvants. Reports reveal that various types of liposomes boost the depot effect in the host body [26].

8.3.2 Enhanced Presentation of Antigen

The coupling of antigen and adjuvant molecules often results in alteration in the physical properties of antigen that promotes slow release of circulating antigen improving the time of activity. It also enhances the surface area of antigen facilitating improved interaction with immunogen. The adjuvants efficaciously transport the antigen to the APCs [26].

8.3.3 Modulation of Immune Response

Adjuvants have a significant impact on immune response regulation through increasing antibody titers, generating distinct types of antibodies, or raising the likelihood of delayed-type allergies. Adjuvants can induce cytokine (Interferon-IFN, IFN-γ IL-2, and IL-12) production to mediate innate and adaptive immune response. Some adjuvants are identified to boost the level neutrophils, dendritic cells, macrophages, and monocytes. The adjuvants activate uptake by cells and processing aiding antigen loading by MHC class I as well as CD8$^+$ cytolytic T-cell priming reaction. This ability of vaccine facilitates elimination of intracellular pathogens [26].

The activation of innate immunity initiates proinflammatory reaction locally, consequently increasing the deployment of immune cells at the injection site. The complex of adjuvant and antigen triggers PRR by presenting as pathogen-associated molecular patterns. As a result, cytokines and chemokines are produced in addition to induction of cells with innate immunity to the site of injection [27]. The loading and presentation of the antigen on the APC is enhanced. The APC further gets activated and matured by raised expression of MHC class II and co-stimulatory molecules stimulating the activation of naïve T cells, resulting in an intensified adaptive immunological response. The fluid is then drained in the lymph nodes and inflammasomes are activated [28].

An overview of the mechanism of vaccine adjuvant is presented in Figure 8.1.

The immune cells exhibit a considerable increase in the gene expression when exposed to alum, CpG-ODN, and MF59 adjuvants. These adjuvants also can influence the expression of these genes. Activating these central response genes stimulates the production of cytokines and chemokines [29]. Aluminum has a crucial function in improving the process of antigen internalization by the antigen-presenting cells. Also, it helps in reducing the pace at which the internalized antigen is degraded [30]. Saponins facilitate the delivery of antigens through the endogenous antigen presentation

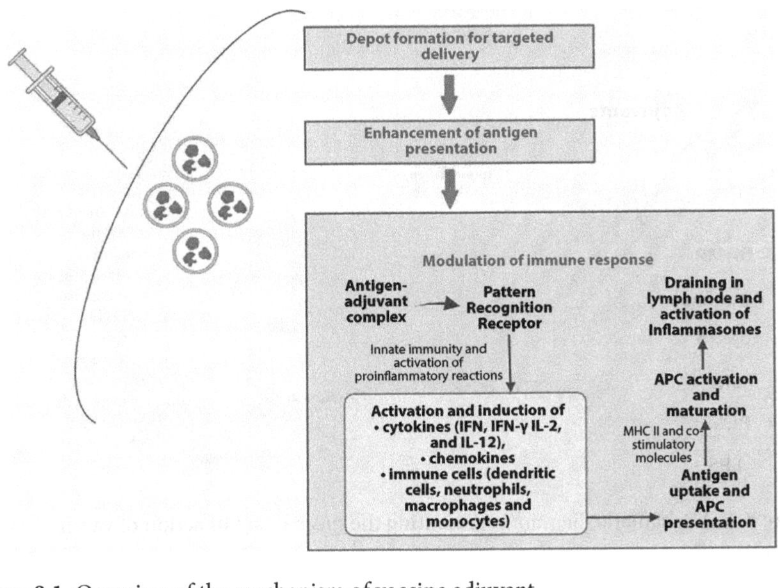

Figure 8.1 Overview of the mechanism of vaccine adjuvant.

system and by augmenting immunological responses mediated by cytotoxic T lymphocytes [31].

TLR agonists used as adjuvants have demonstrated encouraging outcomes by stimulating TLRs, which, in turn, enhance the presentation of antigens, enhance the expression of cytokines and co-stimulatory signals, which will ultimately lead to more adaptive immune responses [32]. Toll-like receptors, present on the outer surface of cells, primarily detect and bind to various components of microbial membranes. Consequently, when immunostimulants activate TLRs on antigen-presenting cells, they often synthesise pro-inflammatory cytokines, including interleukines, interferons, and tumor necrosis factors. This, in turn, results in the development of immune responses of either Th1 or Th2 type [33, 34]. Stimulating these intracellular TLRs by the relevant categories of immune-stimulating substances typically generates type I interferons. Moreover, these type I interferons can promote antigen presentation, Th1 cell differentiation, and regulatory function [35]. In addition, the activation of PRRs such as retinoic acid-induced gene I, melanoma differentiation-associated gene 5, NOD-like receptor thermal protein domain associated protein 3, and nucleotide-binding oligomerization domain 1 and 2 by adjuvants will increase major histocompatibility complex II expression by APCs, thereby enhancing antigen presentation [36]. A schematic diagram representing the mechanism of action of vaccine adjuvants (Figure 8.2).

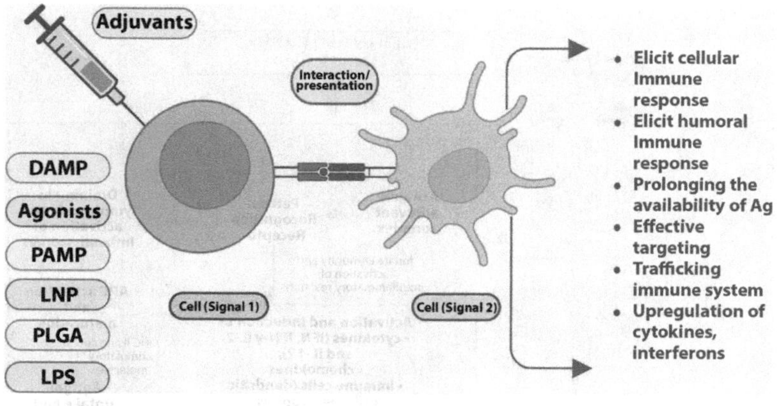

Figure 8.2 A schematic diagram representing the mechanism of action of vaccine adjuvants.

8.4 Adjuvant Platforms

Aluminum adjuvants: Aluminum is extensively utilized as a vaccination adjuvant and has been established as a neurotoxin [37]. Adjuvants composed of aluminum have been used to increase the immunogenicity of vaccines for over a decade. The initial aluminum-adjuvanted vaccines were produced by mixing a base with a mixture of the antigen and aluminum potassium sulfate, according to HogenEsch *et al.* The process described above resulted in the development a precipitate comprising the antigen and aluminum salt [38]. Aluminum salts often referred to as alum consist of aluminum phosphate, amorphous aluminum hydroxy phosphate sulfate, aluminum potassium sulphate, aluminum hydroxide, and aluminum phosphate. It is hypothesized that the adjuvant properties of aluminum are generated through the stimulation of Th2-type immune responses and the facilitation of antibody formation via the activation of B cells, complement system, and immune cell recruitment to the injection site. Moreover, research has indicated that aluminum promotes the maturation of dendritic cells, which are specialized immune cells, and facilitates the intake of antigens [39].

Saponin-based adjuvants: Extensive research has been conducted to examine the adjuvant properties of saponins. Saponins are naturally occurring compounds originating from plants, possessing diverse biological properties. These substances are identified by one or more oligosaccharide chains encircling a lipophilic triterpenoid core. Saponins, which are sourced from plants, have attracted a lot of interest as adjuvants for vaccines because of their various biological and immunomodulatory properties [40]. Chen *et al.* reported that saponins' adjuvant action can be enhanced by the presence of branched sugar chains, aldehyde groups, and an acyl residue on the aglycone [41]. QS-21 is a bioactive component derived from the bark of the *Quillaja saponaria* tree, native to Chile. It is obtained through a purification process involving reverse-phase chromatography. QS-21 as well as the refined portion Quil A saponins are frequently employed as adjuvants due to their remarkable capacity to boost immune responses from antibodies, activate T helper type 1 cells, and stimulate dendritic cells, which in turn activates cytotoxic T-lymphocyte immunological responses [41, 42].

GalCer-based adjuvants: GalCer is a synthetic glycolipid produced from marine sponges. It possesses both antitumor and immunostimulatory properties. α-Galactosylceramide consists of galactose that is α-linked to a ceramide molecule made up of a C26:0 acyl chain and an 18-carbon phytosphingosine chain [43]. αGC is a highly effective T-cell antigen extensively studied in various T lymphocytes. Invariant natural killer T (iNKT) cells, in particular, respond rapidly to αGC by producing a substantial quantity of cytokines [44]. Chemical bonds have been established between α-GalCer alterations and peptide and carbohydrate epitopes to create synthetic vaccinations that combine adjuvant and antigen characteristics in one molecule [45].

Lipopolysaccharide-based adjuvants: A significant portion of gram-negative bacteria outer membrane comprises LPSs, often called endotoxins. It acts as a stimulant for both humoral and cellular responses in humans and can potentially be used as an adjuvant in vaccines [46]. This adjuvant activity is possible after immunization with bacteria-derived vaccines that inherently comprise LPS, as well as by incorporating LPS or its derivative molecules into purified vaccine antigens [47]. It is possible to reduce LPS's toxicity without sacrificing its ability to generate the appropriate immune response for a vaccination meant to prevent a specific disease. Lipid A and LPS, two possible adjuvants, were prepared as an oil-in-water emulsion (OIWE) and tested in BALB/c female mice to determine whether they might generate anti-TNP IgG (anti-2,4,6-trinitrophenyl immunoglobulin G) against the TNP-HEA (hen egg albumin) conjugate. According to Masoud, the activities of the OIWE formulations of all adjuvant candidates based on LPS and lipid A were greater than those of the aqueous formulations [48].

Zwitterionic polysaccharide adjuvants: Zwitterionic polysaccharides (ZPSs) constitute a structurally and biologically distinct category of carbohydrates. Zwitterionic polysaccharides are bacterial polysaccharides that elicit an immune response. They consist of a dense assembly of carbohydrate residues that are both positively and negatively charged. PS A1, PS B, and PS A2 are the preeminent ZPSs derived from a strain of *Bacteroides fragilis*. ZPSs have been documented to elicit stimulation of the innate immune system via interaction with TLR2, thus establishing a connection between the innate and adaptive immune systems. There is evidence that the PS A1 polysaccharide can take on a secondary structure in order to correctly orient the positively and negatively charged groups in order to bind with MHC-II molecules and present the ZPS to its T-cell receptors [49, 50]. Furthermore, numerous synthetic endeavors have been made to ZPSs within the chemical community. The chemical synthesis of zwitterionic oligosaccharide repeating units, which include Sp1, PS A1, CP5, and CP8, was one such attempt [50].

Non-aluminum adjuvants: The MF59 adjuvant functions as a non-aluminum–based adjuvant. When combined with an antigen, it can work as an emulsion delivery system that gradually releases antigens into the lymph nodes, extending the length of antigen contact with the immune system and enhancing antigen presentation [51]. When MF59 is injected into the muscle, it stimulates innate immune cells and promotes the synthesis of chemokines. These chemokines subsequently attract additional innate immune cells to the injection site, thereby enhancing the immunological responses. They also facilitate the movement of these acquired innate immune cells to the nearby lymph nodes, stimulating the activation of T and B cells [19].

TLR agonist molecule-based adjuvants: AS04 and CpG ODN 1018 are adjuvants based on classical TLR agonist molecules. Toll-like receptor agonists show great potential as adjuvants in vaccines targeting severe and intricate diseases, including HIV AIDS, Malaria, and Cancer [33]. The TLR encodes a leucine-rich repeat segment that facilitates the recognition of PAMP/DAMP and a toll-interleukin-1 receptor (TIR) domain that initiates an inflammatory response and transmits the subsequent signal transduction [52]. CpG oligodeoxynucleotides can attach to and stimulate TLRs, which triggers an innate immune response that facilitates the eventual generation of adaptive immunity. Moreover, CpG ODNs exhibit significant potential for incorporating into preventive vaccinations to prevent infection [53].

Other adjuvant derivatives: Synthetic double-stranded RNAs have the ability to target and activate various TLRs found on antigen-presenting cells. This activation results in the generation of pro-inflammatory cytokines such as interleukins and interferons. Furthermore, it increases the development of potent Th1-biased immune responses and cytotoxic T cells [54]. When used as an adjuvant for the H5N1 influenza vaccination, Glucopyranosyl lipid adjuvant-stable emulsion (GLA-SE) has been proven to reduce the required dose. Multiple clinical trials investigated the use of GLA-SE as an adjuvant for the tuberculosis vaccination alongside the ID93 recombinant protein. The results revealed that the vaccine efficiently promoted the generation of particular antibodies and CD4+T responses, indicating that it was safe and capable of eliciting an immune response in the vaccinated individuals [55].

8.5 Regulatory Guidelines for Vaccine Adjuvants

In recent years, there has been an increase in interest in vaccination adjuvants and innovative delivery techniques [56]. The subject of vaccine development is constantly evolving as new vaccines are researched and developed for many infectious diseases, including those that cause infectious diarrhea. An adjuvant

Table 8.1 Regulatory bodies and their approved vaccines.

Sl. no.	Regulatory bodies	Guidelines proposed for on vaccine adjuvants	Example of approved vaccines with adjuvants	References
1	World Health Organization (WHO)	It furnishes fundamental principles and guidelines for the evaluation of vaccine safety, with a focus on considerations related to adjuvants.	BBV152/Covaxin (alum-imidazoquinolin-adjuvanted vaccine)	[73, 74]
2	U.S. Food and Drug Administration (FDA)	FDA describes high standards for adjuvant effectiveness and safety, highlighting the need of well-conducted clinical studies.	Hepatitis B vaccine (contains aluminum salts as an adjuvant)	[75]
3	European Medicines Agency (EMA)	EMA provides recommendations for assessing vaccinations, taking adjuvants factors into consideration to guarantee safety and effectiveness.	ADACEL for tetanus toxoid, reduced diphtheria toxoid, and acellular pertussis vaccine (contains aluminum phosphate as an adjuvant)	[76]
4	Health Canada	It specifies criteria for the approval of vaccinations and the adjuvants that are employed in them under Canadian jurisdiction, with a focus on safety and effectiveness evaluations.	Shingrix (herpes zoster vaccine with an adjuvant called AS01B)	[77]

(Continued)

Table 8.1 Regulatory bodies and their approved vaccines. (*Continued*)

Sl. no.	Regulatory bodies	Guidelines proposed for on vaccine adjuvants	Example of approved vaccines with adjuvants	References
5	Australian Therapeutic Goods Administration (TGA)	TGA lays forth legal requirements for adjuvants in vaccinations that are licensed for use in Australia, ensuring both their safety and effectiveness.	Fluad (seasonal influenza vaccine containing MF59 adjuvant)	[78]

is meant to boost the immunological response to an antigen found in a vaccine. It is of the utmost importance to give a justification for the use of an adjuvant that is substantiated by data in an appropriate animal model, especially in light of the regulatory guidelines and scrutiny from the public that concerns its application. Adjuvants must be evaluated specifically for immunogenicity and toxicology because not all adjuvants are intended to trigger the same kind or prolonged immune system activation [57]. The safety, effectiveness, and overall efficacy of vaccines containing adjuvants are guaranteed by regulatory bodies and health authorities through the establishment of regulatory guidelines. The regulation of adjuvant vaccines involves the review of each stage of the production process, compliance with GMPs, and post-marketing surveillance or pharmacovigilance. The authorization bodies focus on toxicity tests, immunogenicity studies, biodistribution analysis and regulatory submission prior to initiating the clinical trials. Details of guidelines for vaccines with adjuvants approved by regulatory bodies are presented in Table 8.1.

8.5.1 World Health Organization

The World Health Organization (WHO) regulatory requirements for vaccine adjuvants are essential for guaranteeing the effectiveness and safety of vaccinations. The WHO offers broad recommendations and standards for assessing vaccination safety, taking adjuvants into account, through its Global vaccination Safety Initiative. A number of national regulatory bodies, including Health Canada, the EMA, and the U.S. FDA, provide specific standards for the licensing of vaccines and their adjuvants within their respective regions [58]. WHO provides scientific and recommendatory guidelines for the manufacturers and national regulatory authorities (NRAs). According to the WHO guidelines, the antigen, adjuvant and final vaccine products should be characterized with toxicological studies. Safety of the subjects to whom the vaccine is to be administered should be assured by the manufacturers. Consequently, various documentation and certifications are required during the course of production and testing to obtain approval from regulatory bodies.

The adjuvanted vaccine manufactured should comply with the Good Manufacturing Practices during the clinical stages as proposed by the WHO guidelines. Quality control evaluations should be conducted and refined frequently. In case of absence of standards, the manufacturers are expected to provide a primary reference material during the clinical stages. Together, these regulatory rules seek to guarantee that vaccines and adjuvants receive comprehensive review, striking a balance between the strict safety requirements and a need for efficacy, therefore increasing trust in vaccination programs throughout the globe.

8.5.2 Food and Drug Administration

The United States FDA implements rigorous rules and regulations for vaccine adjuvants in order to validate their safety as well as effectiveness in vaccine production [59]. The investigation method includes a thorough examination of adjuvant safety, emphasizing the importance of comprehensive evaluation through well-controlled clinical studies. In addition to a thorough examination of probable systemic and local reactions, safety considerations also include an evaluation of immunological responses and potential long-term consequences [60]. The FDA licenses adjuvant vaccines as combination of adjuvants with particular antigens. The regulatory authorization is a product-based approach. As vaccines are categorized as biologics, the regulatory guidelines are presented in Title 21 of the Code of Federal Regulations (21 C.F.R.) of Section 610. The rules on testing potency, lot release, and efficacy are provided in the section. Prior to transitioning to human trials, it is crucial to conduct preclinical studies, frequently employing animal models, to evaluate safety and immunogenicity and identify potential adverse effects. A fundamental aspect of this process involves well-regulated clinical trials, incorporating a diverse range of human populations to comprehensively assess vaccine performance across different demographic groups. The FDA, for instance, sets strict guidelines for adjuvant safety and effectiveness, highlighting the necessity of carefully monitored clinical studies. The hepatitis B vaccine is one of the approved vaccinations that makes use of adjuvants; it frequently includes aluminum salts as an adjuvant [27].

The FDA carefully evaluates the risk-benefit balance of vaccinations that contained adjuvants, to guarantee any possible threats are mitigated by the accompanying benefits, particularly the vaccine's increased efficacy. Emphasizing continuous post-marketing surveillance, there is a vigilant monitoring of adjuvanted vaccines to detect any uncommon or delayed adverse events. This ongoing surveillance plays a critical role in frequently evaluating the safety and effectiveness of these vaccines. Examples of adjuvanted vaccines granted FDA approval encompass hepatitis B vaccines like Engerix-B and Recombivax HB, potentially incorporating aluminum salts as adjuvants. Additionally, the human papillomavirus (HPV) vaccine Gardasil 9 has secured approval, featuring an adjuvant to enhance the immune response [61–63].

Furthermore, the FDA emphasizes ongoing information exchange and cooperation between funders and regulatory organizations to facilitate the research and regulatory approval procedures for vaccines using adjuvants [64]. As scientific knowledge and technology continue to advance, the regulatory context undergoes transformations, prompting adjustments in FDA guidelines. Evolving with ongoing research, the FDA may refine its recommendations to stay abreast of the latest scientific developments and effectively address emerging challenges.

8.5.3 European Medicines Agency

Adjuvant vaccines are subject to strict regulatory standards from the EMA in order to ensure their safety and effectiveness during the development and licensing process [65]. These recommendations provide a strong emphasis on a number of important subjects, such as adjuvant quality and composition, which must meet stringent requirements and have a clear and standardized composition. The production process is closely monitored to guarantee repeatability and compliance with good manufacturing practices. According to the EMA, the advantages of adjuvants in enhancing the immune response without causing the development of systemic or local adverse events should be validated with evidence. Animal studies and other non-clinical research are vital for determining the safety profile of adjuvanted vaccinations, pointing up possible toxicological issues, and directing the planning of subsequent human trials. Clinical development requires undertaking well-designed and controlled trials to thoroughly investigate the safety, immunogenicity, and efficacy of adjuvanted vaccines in a wide range of human populations. The EMA undertakes a complete risk-benefit evaluation, assessing possible hazards against the advantages of increased immune responses, with an emphasis on explaining any observed concerns by demonstrated benefits. Post-approval surveillance is still essential for the continuing safety and efficacy evaluations for adjuvanted vaccinations. Adjuvanted vaccinations that have been approved by the EMA include HPV vaccines, such as Cervarix and Gardasil 9, which are intended to combat specific infectious risks and improve immunogenicity [66].

- Human Papillomavirus (HPV) Vaccines

Cervarix (GlaxoSmithKline): This HPV vaccination targets various types of the virus that have been linked to cervical cancer. It comprises the AS04 adjuvant system, which includes aluminum hydroxide and monophosphoryl lipid A. The adjuvant increases immunological reaction, which raises the effectiveness of the vaccination [67].

Gardasil 9 (Merck): This HPV vaccination protects against a larger range of HPV types linked to cervical and other malignancies. It also has an adjuvant system that aids in the activation of a strong immunological response [67].

- Pandemic Influenza Vaccines

Focetria (Novartis, now Seqirus): Focetria (Novartis, now Seqirus) is an influenza vaccine that contains the adjuvant MF59, which was created in response to the H1N1 pandemic. The oil-in-water emulsion MF59 strengthens the body's defenses against immunization antigens, hence enhancing the effectiveness of the vaccination [69].

Pandemrix (GlaxoSmithKline): Another influenza vaccine developed during the H1N1 pandemic is called Pandemrix; it also includes the AS03 adjuvant system. The purpose of AS03 is to increase the immunogenicity of the vaccine by including polysorbate 80, squalene, and alpha-tocopherol (vitamin E) [19].

These examples demonstrate the use of adjuvants in vaccinations that have been authorized by the EMA.

8.5.4 Health Canada

In Canada, Health Canada is instrumental in shaping regulatory guidelines for vaccine adjuvants, with the overarching goal of ensuring the efficacy, safety, and quality of vaccines [69]. Aligned with global standards, these guidelines underscore a meticulous evaluation process. Adjuvants are subject to rigorous scrutiny to fulfill precise quality benchmarks, requiring well-defined and consistent compositions.

Health Canada has granted approval to adjuvanted vaccines such as Shingrix, designed for herpes zoster and featuring the AS01B adjuvant, and Fluad, a seasonal influenza vaccine incorporating the MF59 adjuvant [70]. These regulatory guidelines are instrumental in ensuring that vaccines with adjuvants not only meet but exceed the stringent standards for safety and effectiveness set within the Canadian healthcare framework.

8.5.5 Australian Therapeutic Goods Administration (TGA)

The Therapeutic Goods Administration (TGA) in Australia establishes regulatory requirements for vaccination adjuvants to assure their safety, effectiveness, and quality. These guidelines highlight a thorough examination procedure and are in accordance with international norms [71]. Additionally, TGA establishes a strong emphasis on risk-benefit analysis, which balances possible hazards with the advantages of enhanced vaccination efficiency.

The seasonal influenza vaccination Fluad, which contains the MF59 adjuvant to boost the immune response, is one example of an adjuvanted vaccine that has been authorized by the TGA [72]. The purpose of these regulations is to provide assurance that adjuvanted vaccinations in the Australian setting adhere to the utmost safety and efficacy guidelines.

8.6 Conclusion

Finally, regulatory specifications for vaccine adjuvants are critical in guaranteeing vaccine safety, efficacy, and quality. These potential recommendations may provide a complete framework for developing, testing, and approving adjuvanted vaccines, addressing essential issues such as formulation, manufacturing, and clinical evaluation. To protect public health, certain regulatory standards must be met in order to reduce hazards and maximize advantages of vaccine adjuvants. As science progresses and new adjuvant technologies emerge, constant collaboration among regulatory authorities, researchers, and manufacturers is essential for adapting guidelines to probably increase scientific understanding. Furthermore, a clear regulatory framework supports the worldwide fight against infectious diseases by fostering innovation in vaccine research and fostering public confidence in the efficacy and safety of adjuvanted vaccinations.

Acknowledgments

The authors would like to express our sincere gratitude to Dr. G. Viswanathan, Chancellor, Vellore Institute of Technology, Vellore for providing encouragement and support.

References

1. Rodrigues, C.M.C. and Plotkin, S.A., Impact of Vaccines; Health, Economic and Social Perspectives. *Front. Microbiol.*, 11, 1526, 2020.
2. Mortellaro, A. and Ricciardi-Castagnoli, P., From vaccine practice to vaccine science: the contribution of human immunology to the prevention of infectious disease. *Immunol. Cell Biol.*, 89, 3, 332–339, 2011.
3. Herati, R.S. and Wherry, E.J., What Is the Predictive Value of Animal Models for Vaccine Efficacy in Humans? *Cold Spring Harb. Perspect. Biol.*, 10, 4, a031583, 2018.
4. Chavda, V.P., Balar, P.C., Apostolopoulos, V., History of vaccination, in: *Advanced Vaccination Technologies for Infectious and Chronic Diseases*, pp. 1–12, Elsevier, 2024, [Internet] [cited 2024 Apr 25]. Available from: https://linkinghub.elsevier.com/retrieve/pii/B978044318564900028X.
5. Chaudhary, N., Weissman, D., Whitehead, K.A., mRNA vaccines for infectious diseases: principles, delivery and clinical translation. *Nat. Rev. Drug Discov.*, 20, 11, 817–838, 2021.
6. Chavda, V.P., Balar, P.C., Jogi, G. *et al.*, The potential role of essential oils in boosting immunity and easing COVID-19 symptoms. *Clin. Tradit. Med. Pharmacol.*, 5, 1, 200135, 2024.
7. Chavda, V.P., Vuppu, S., Mishra, T., Balar, P., The Emergence of Hybrid Variants of SARS-CoV-2: Towards Hybrid Immunity. *Vaccines*, 11, 4, 764, 2023.
8. Chavda, V.P., Mishra, T., Kamaraj, S. *et al.*, Post-COVID-19 Fungal Infection in the Aged Population. *Vaccines*, 11, 3, 555, 2023.
9. Brisse, M., Vrba, S.M., Kirk, N., Liang, Y., Ly, H., Emerging Concepts and Technologies in Vaccine Development. *Front. Immunol.*, 11, 583077, 2020.
10. Umscheid, C.A., Margolis, D.J., Grossman, C.E., Key Concepts of Clinical Trials: A Narrative Review. *Postgrad. Med.*, 123, 5, 194–204, 2011.
11. Lythgoe, M.P. and Middleton, P., Comparison of COVID-19 Vaccine Approvals at the US Food and Drug Administration, European Medicines Agency, and Health Canada. *JAMA Netw. Open*, 4, 6, e2114531, 2021.
12. Rezaie, R., McGahan, A.M., Frew, S.E., Daar, A.S., Singer, P.A., Emergence of biopharmaceutical innovators in China, India, Brazil, and South Africa as global competitors and collaborators. *Health Res. Policy Syst.*, 10, 1, 18, 2012.
13. Tang, L., Meissner, E.G., Kottilil, S., 28 - Virology and Pathogenesis of Hepatitis C, in: *Zakim and Boyer's Hepatology*, Seventh Edition, vol. e8, A.J. Sanyal, T.D. Boyer, K.D. Lindor, N.A. Terrault (Eds.), pp. 412–427, Elsevier, Philadelphia, 2018, [Internet] [cited 2023 Dec 2]. Available from: https://www.sciencedirect.com/science/article/pii/B9780323375917000288.
14. Chevalier-Cottin, E.-P., Ashbaugh, H., Brooke, N. *et al.*, Communicating Benefits from Vaccines Beyond Preventing Infectious Diseases. *Infect. Dis. Ther.*, 9, 3, 467–480, 2020.

15. Betsch, C., Böhm, R., Korn, L., Holtmann, C., On the benefits of explaining herd immunity in vaccine advocacy. *Nat. Hum. Behav.*, 1, 3, 1–6, 2017.
16. Saleh, A., Qamar, S., Tekin, A., Singh, R., Kashyap, R., Vaccine Development Throughout History. *Cureus*, 13, 7, e16635.
17. Zhao, T., Cai, Y., Jiang, Y. *et al.*, Vaccine adjuvants: mechanisms and platforms. *Signal Transduct. Target. Ther.*, 8, 1, 1–24, 2023.
18. Reed, S.G., Orr, M.T., Fox, C.B., Key roles of adjuvants in modern vaccines. *Nat. Med.*, 19, 12, 1597–1608, 2013.
19. Pulendran, B., Arunachalam P, S., O'Hagan, D.T., Emerging concepts in the science of vaccine adjuvants. *Nat. Rev. Drug Discov.*, 20, 6, 454–475, 2021.
20. Didierlaurent, A.M., Laupèze, B., Di Pasquale, A., Hergli, N., Collignon, C., Garçon, N., Adjuvant system AS01: helping to overcome the challenges of modern vaccines. *Expert Rev. Vaccines*, 16, 1, 55–63, 2017.
21. Zhou, C., Xu, Z., Li, G., Gao, Q., Sui, Q., Li, T., Efficient synthesis of monophosphoryl lipid A mimetic RC-529. *J. Carbohydr. Chem.*, 40, 7–9, 501–517, 2021.
22. Stoute, J.A., Slaoui, M., Heppner, D.G. *et al.*, A preliminary evaluation of a recombinant circumsporozoite protein vaccine against Plasmodium falciparum malaria. RTS,S Malaria Vaccine Evaluation Group. *N. Engl. J. Med.*, 336, 2, 86–91, 1997.
23. Leroux-Roels, I., Borkowski, A., Vanwolleghem, T. *et al.*, Antigen sparing and cross-reactive immunity with an adjuvanted rH5N1 prototype pandemic influenza vaccine: a randomised controlled trial. *Lancet Lond. Engl.*, 370, 9587, 580–589, 2007.
24. Einstein, M.H., Baron, M., Levin, M.J. *et al.*, Comparison of the immunogenicity and safety of Cervarix and Gardasil human papillomavirus (HPV) cervical cancer vaccines in healthy women aged 18-45 years. *Hum. Vaccin.*, 5, 10, 705–719, 2009.
25. Bourne, N., Bravo, F.J., Francotte, M. *et al.*, Herpes simplex virus (HSV) type 2 glycoprotein D subunit vaccines and protection against genital HSV-1 or HSV-2 disease in guinea pigs. *J. Infect. Dis.*, 187, 4, 542–549, 2003.
26. Wu, Z. and Liu, K., Overview of vaccine adjuvants. *Med. Drug Discov.*, 11, 100103, 2021.
27. Facciolà, A., Visalli, G., Laganà, A., Di Pietro, A., An Overview of Vaccine Adjuvants: Current Evidence and Future Perspectives. *Vaccines*, 10, 5, 819, 2022.
28. Awate, S., Babiuk, L.A., Mutwiri, G., Mechanisms of Action of Adjuvants. *Front. Immunol.*, 4, 114, 2013.
29. Mosca, F., Tritto, E., Muzzi, A. *et al.*, Molecular and cellular signatures of human vaccine adjuvants. *Proc. Natl. Acad. Sci. U.S.A.*, 105, 30, 10501–10506, 2008.
30. Ghimire, T.R., Benson, R.A., Garside, P., Brewer, J.M., Alum increases antigen uptake, reduces antigen degradation and sustains antigen presentation by DCs *in vitro*. *Immunol. Lett.*, 147, 1–2, 55–62, 2012.

31. Sjölander, A., Drane, D., Maraskovsky, E. *et al.*, Immune responses to ISCOM® formulations in animal and primate models. *Vaccine*, 19, 17, 2661–2665, 2001.
32. Duthie, M.S., Windish, H.P., Fox, C.B., Reed, S.G., Use of defined TLR ligands as adjuvants within human vaccines. *Immunol. Rev.*, 239, 1, 178–196, 2011.
33. Luchner, M., Reinke, S., Milicic, A., TLR Agonists as Vaccine Adjuvants Targeting Cancer and Infectious Diseases. *Pharmaceutics*, 13, 2, 142, 2021.
34. Miyake, K., Innate immune sensing of pathogens and danger signals by cell surface Toll-like receptors. *Semin. Immunol.*, 19, 1, 3–10, 2007.
35. Aleynick, M., Svensson-Arvelund, J., Flowers, C.R., Marabelle, A., Brody, J.D., Pathogen Molecular Pattern Receptor Agonists: Treating Cancer by Mimicking Infection. *Clin. Cancer Res. Off. J. Am. Assoc. Cancer Res.*, 25, 21, 6283–6294, 2019.
36. Maisonneuve, C., Bertholet, S., Philpott, D.J., De Gregorio, E., Unleashing the potential of NOD- and Toll-like agonists as vaccine adjuvants. *Proc. Natl. Acad. Sci. U.S.A.*, 111, 34, 12294–12299, 2014.
37. T, L. and Ca, S., Aluminum vaccine adjuvants: are they safe? *Curr. Med. Chem.*, 18, 17, 2630–2637, 2011, [Internet]Available from: https://pubmed.ncbi.nlm.nih.gov/21568886/.
38. HogenEsch, H., O'Hagan, D.T., Fox, C.B., Optimizing the utilization of aluminum adjuvants in vaccines: you might just get what you want. *npj Vaccines*, 3, 1, 1–11, 2018.
39. Djurisic, S., Jakobsen, J.C., Petersen, S.B., Kenfelt, M., Klingenberg, S.L., Gluud, C., Aluminium adjuvants used in vaccines. *Cochrane Database Syst. Rev.*, 2018, 7, CD013086, 2018.
40. Sun, H.-X., Xie, Y., Ye, Y.-P., Advances in saponin-based adjuvants. *Vaccine*, 27, 12, 1787–1796, 2009.
41. Chen, K., Wang, N., Zhang, X., Wang, M., Liu, Y., Shi, Y., Potentials of saponins-based adjuvants for nasal vaccines. *Front. Immunol.*, 14, 1–14, 2023, [Internet] Available from: https://www.frontiersin.org/articles/10.3389/fimmu.2023.1153042.
42. Zhu, D. and Tuo, W., QS-21: A Potent Vaccine Adjuvant. *Nat. Prod. Chem. Res.*, 3, 4, e113, 2016.
43. Garcia-Vello, P., Speciale, I., Chiodo, F., Molinaro, A., De Castro, C., Carbohydrate-based adjuvants. *Drug Discov. Today Technol.*, 35–36, 57–68, 2020.
44. Zhao, M. and Kronenberg, M., Lymphocytes That Participate in Innate Immune Responses, in: *Reference Module in Biomedical Sciences*, Elsevier, 2014, [Internet] [cited 2023 Nov 27]. Available from: https://www.sciencedirect.com/science/article/pii/B9780128012383001070.
45. Abdel-Aal, A.-B.M., El-Naggar, D., Zaman, M., Batzloff, M., Toth, I., Design of Fully Synthetic, Self-Adjuvanting Vaccine Incorporating the

Tumor-Associated Carbohydrate Tn Antigen and Lipoamino Acid-Based Toll-like Receptor 2 Ligand. *J. Med. Chem.*, 55, 15, 6968–6974, 2012.

46. Arenas, J., Bacterial Lipopolysaccharide as Adjuvants, in: *Molecular Vaccines: From Prophylaxis to Therapy*, vol. 2, M. Giese (Ed.), pp. 527–536, Springer International Publishing, Cham, 2014, [Internet] [cited 2023 Nov 27]. Available from: https://doi.org/10.1007/978-3-319-00978-0_7.
47. Zariri, A. and van der Ley, P., Biosynthetically engineered lipopolysaccharide as vaccine adjuvant. *Expert Rev. Vaccines*, 14, 6, 861–876, 2015.
48. Masoud, H., Novel adjuvants derived from attenuated lipopolysaccharides and lipid As of purple non-sulfur photosynthetic bacteria. *Vaccine*, 37, 26, 3472–3477, 2019.
49. Wang, Z., Poveda, A., Zhang, Q. *et al.*, Total Synthesis and Structural Studies of Zwitterionic Bacteroides fragilis Polysaccharide A1 Fragments. *J. Am. Chem. Soc.*, 145, 25, 14052–14063, 2023.
50. Zhang, Q., Overkleeft, H.S., van der Marel, G.A., Codée, J.D., Synthetic zwitterionic polysaccharides. *Curr. Opin. Chem. Biol.*, 40, 95–101, 2017.
51. Cantisani, R., Pezzicoli, A., Cioncada, R. *et al.*, Vaccine adjuvant MF59 promotes retention of unprocessed antigen in lymph node macrophage compartments and follicular dendritic cells. *J. Immunol. Baltim. Md 1950*, 194, 4, 1717–1725, 2015.
52. Yang, J.-X., Tseng, J.-C., Yu, G.-Y. *et al.*, Recent Advances in the Development of Toll-like Receptor Agonist-Based Vaccine Adjuvants for Infectious Diseases. *Pharmaceutics*, 14, 2, 423, 2022.
53. Shirota, H. and Klinman, D.M., Chapter 9 - CpG Oligodeoxynucleotides as Adjuvants for Clinical Use, in: *Immunopotentiators in Modern Vaccines*, Second Edition, V.E.J.C. Schijns and D.T. O'Hagan (Eds.), pp. 163–198, Academic Press, 2017, [Internet] [cited 2023 Nov 27]. Available from: https://www.sciencedirect.com/science/article/pii/B9780128040195000098.
54. Coler, R.N., Day, T.A., Ellis, R. *et al.*, The TLR-4 agonist adjuvant, GLA-SE, improves magnitude and quality of immune responses elicited by the ID93 tuberculosis vaccine: first-in-human trial. *npj Vaccines*, 3, 34, 2018.
55. Pillet, S., Aubin, E., Trépanier, S. *et al.*, Humoral and cell-mediated immune responses to H5N1 plant-made virus-like particle vaccine are differentially impacted by alum and GLA-SE adjuvants in a Phase 2 clinical trial. *npj Vaccines*, 3, 1, 1–9, 2018.
56. Sesardic, D. and Dobbelaer, R., European union regulatory developments for new vaccine adjuvants and delivery systems. *Vaccine*, 22, 19, 2452–2456, 2004.
57. Plitnick, L.M., Global Regulatory Guidelines for Vaccines. *Nonclinical Dev. Nov. Biol. Biosimilars Vaccines Spec. Biol.*, 225–241, 2013.
58. Da Silva, F.T., Di Pasquale, A., Yarzabal, J.P., Garçon, N., Safety assessment of adjuvanted vaccines: Methodological considerations. *Hum. Vaccines Immunother.*, 11, 7, 1814–1824, 2015.
59. Mohan, T., Verma, P., Rao, D.N., Novel adjuvants & delivery vehicles for vaccines development: A road ahead. *Indian J. Med. Res.*, 138, 5, 779–795, 2013.

60. Garçon, N., Segal, L., Tavares, F., Van Mechelen, M., The safety evaluation of adjuvants during vaccine development: the AS04 experience. *Vaccine*, 29, 27, 4453–4459, 2011.
61. Nooraei, S., Bahrulolum, H., Hoseini, Z.S. *et al.*, Virus-like particles: preparation, immunogenicity and their roles as nanovaccines and drug nanocarriers. *J. Nanobiotechnol.*, 19, 1, 59, 2021.
62. Chavda, V.P., Nalla, L.V., Balar, P. *et al.*, Advanced Phytochemical-Based Nanocarrier Systems for the Treatment of Breast Cancer. *Cancers*, 15, 4, 1023, 2023.
63. Chavda, V.P., Balar, P.C., Dodiya, P., Bojarska, J., Nanocarrier-based delivery of peptides: challenges and way forward. *Nanomed.*, 18, 26, 1903–1906, 2023.
64. Feigal, E.G., Tsokas, K., Zhang, J., Cromer, M.V., Whittlesey, K.J., Werner, M.J., Perspective: Communications with the Food and Drug Administration on the Development Pathway for a Cell-Based Therapy: Why, What, When, and How? *Stem Cells Transl. Med.*, 1, 11, 825–832, 2012.
65. Wagner, R., Meißner, J., Grabski, E., Sun, Y., Vieths, S., Hildt, E., Regulatory concepts to guide and promote the accelerated but safe clinical development and licensure of COVID-19 vaccines in Europe. *Allergy*, 77, 1, 72–82, 2022.
66. Matić, Z. and Šantak, M., Current view on novel vaccine technologies to combat human infectious diseases. *Appl. Microbiol. Biotechnol.*, 106, 1, 25–56, 2022.
67. Cheng, L., Wang, Y., Du, J., Human Papillomavirus Vaccines: An Updated Review. *Vaccines*, 8, 3, 391, 2020.
68. Tregoning, J.S., Russell, R.F., Kinnear, E., Adjuvanted influenza vaccines. *Hum. Vaccines Immunother.*, 14, 3, 550–564, 2018.
69. MacDonald, N. and Pickering, L., Canada's eight-step vaccine safety program: Vaccine literacy. *Paediatr. Child Health*, 14, 9, 605–608, 2009.
70. Allen, J., Evaluation of Salmonella enterica serovar Typhimurium Vaccines in the Context of Immunosenescence, 2023, Available from: https://archive.hshsl.umaryland.edu/handle/10713/20749.
71. Administration (TGA) TG. COVID-19 vaccines regulatory status | Therapeutic Goods Administration (TGA), 2022, [Internet] Available from: https://www.tga.gov.au/products/covid-19/covid-19-vaccines/covid-19-vaccines-regulatory-status.
72. Banzhoff, A., Pellegrini, M., Del Giudice, G., Fragapane, E., Groth, N., Podda, A., MF59®-adjuvanted vaccines for seasonal and pandemic influenza prophylaxis. *Influenza Other Respir. Viruses*, 2, 6, 243–249, 2008.
73. Newsroom-WHO, The Bharat Biotech BBV152 COVAXIN vaccine against COVID-19: What you need to know, World Health Organization, 2022.
74. Vikkurthi, R., Ansari, A., Pai, A.R. *et al.*, Inactivated whole-virion vaccine BBV152/Covaxin elicits robust cellular immune memory to SARS-CoV-2 and variants of concern. *Nat. Microbiol.*, 7, 7, 974–985, 2022.

75. Wang, Z., Li, S., Shan, P. *et al.*, Improved Aluminum Adjuvants Eliciting Stronger Immune Response When Mixed with Hepatitis B Virus Surface Antigens. *ACS Omega*, 7, 38, 34528–34537, 2022.
76. Mei, C., Deshmukh, S., Cronin, J. *et al.*, Aluminum Phosphate Vaccine Adjuvant: Analysis of Composition and Size Using Off-Line and In-Line Tools. *Comput. Struct. Biotechnol. J.*, 17, 1184–1194, 2019.
77. Syed, Y.Y., Recombinant Zoster Vaccine (Shingrix®): A Review in Herpes Zoster. *Drugs Aging*, 35, 12, 1031–1040, 2018.
78. Li-Kim-Moy, J., Yin, J.K., Rashid, H. *et al.*, Systematic review of fever, febrile convulsions and serious adverse events following administration of inactivated trivalent influenza vaccines in children. *Eurosurveillance*, 20, 24, 21159, 2015.

9

Adjuvant and Vaccine Safety

Shalini Bhattacharya[1], Jyoti Singh[1], Rupesh K. Gautam[1*], Nadeem Farooqui[1], Nimita Manocha[1] and Hitesh Malhotra[2]

[1]*Indore Institute of Pharmacy, IIST Campus, Rau-Indore, Madhya Pradesh, India*
[2]*Guru Gobind Singh College of Pharmacy, Yamunanagar, Haryana, India*

Abstract
In order to increase immunogenicity, adjuvants are substances added to vaccines; in recent decades, research on the synthesis of these molecules has grown significantly. It is a procedure that involves simulating an illness within the body, which ultimately triggers the immune system to produce a strong immunological response. They boost and change the immune response. Recent research links a variety of pathways to the emergence of autoimmune illnesses brought on by vaccination adjuvants, particularly those connected to aluminum-based chemicals commonly called (Alum), which make up the majority of modern adjuvants. Despite new evidence suggesting that vaccinations with adjuvants other than alum may also produce particular autoimmune adverse effects, there are less and less systematic data available concerning other adjuvants. This paper gives information regarding the vaccine-adjuvant classification, characteristic properties of adjuvant, and licensed vaccination using adjuvants and approaches for safety of adjuvant testing.

***Keywords*:** Vaccination, adjuvants, immune system, booster dose, major histocompatibility complex, adjuvant safety study

9.1 Introduction

The human race has benefited from vaccinations against a wide range of infectious illnesses. It is a procedure that involves simulating an illness within the body, which ultimately triggers the immune system to produce

**Corresponding author*: drrupeshgautam@gmail.com

Vivek P. Chavda and Vasso Apostolopoulos (eds.) Emerging Pathways of Vaccine Adjuvants: A Nonspecific Stimulant of the Immune System, (235–254)

a strong immunological response [1]. Infectious illness prevention can be achieved to a great extent by getting vaccination. Adding well-characterized adjuvants that only activate the immune response component necessary for protection—and do not cause a wider widespread stimulation of the immune response—is a typical approach for developing next-generation vaccines.

The utilization of adjuvants plays major role by following functioning that they might work to boost the immune response, lowering the antigen dosages needed for a strong response to immunization while raising the range of response. Furthermore, adjuvants overcome antigenic competition by enabling the creation of complex combination vaccinations.

Furthermore, adjuvants have a critical role in augmenting the restricted immune response observed in some populations, including the elderly, young children, those with chronic illnesses, and immunocompromised people [2].

Adjuvants may be broadly classified into two classes: vehicles and particular immune response enhancers. Vehicles include mineral salts, emulsions, liposomes, and virosomes. They convey vaccination antigens to the immune system more effectively and regulate the release and storage of antigens. Immunostimulants, which influence the immune system and

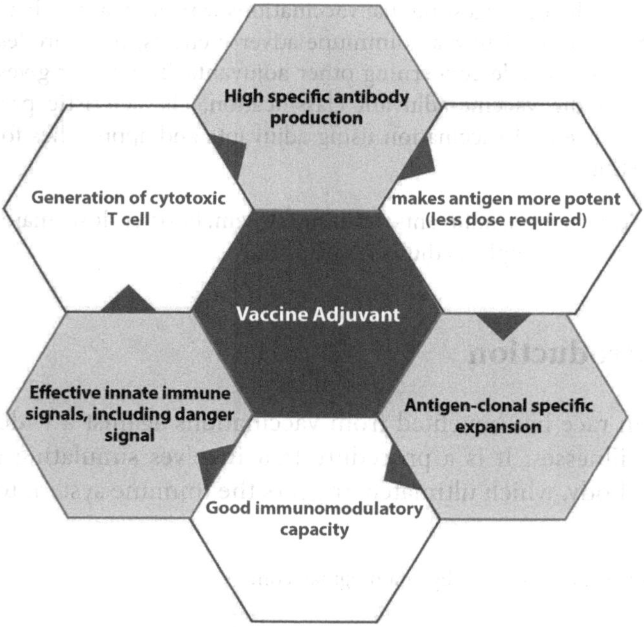

Figure 9.1 Characteristic properties of vaccine adjuvant and its benefits.

heighten the immunological response to antigens, make up the second class of adjuvants [3].

For the majority of the approved inactivated and subunit vaccines, aluminum salts have been shown to be sufficient in eliciting an acceptable immune response. A large number of contemporary vaccinations are made of highly purified proteins without the innate immune system's normal triggers. These improved formulations frequently exhibit decreased immunogenicity, necessitating adjuvant treatment. Using creative adjuvants, such as mixing many adjuvant types into one formulation, is one of the potential ways to increase the effectiveness of recently produced preventive and therapeutic vaccines [4]. Figure 9.1 gives the idea of benefits and characteristic properties of adjuvants when used in vaccines.

9.1.1 Classification of Adjuvants

Adjuvants are, nowadays, majorly experimented whether in combinations or by considering specific health condition or need, and it is divided into various types according to the formulation and functioning or mechanism of action as follows:

9.1.1.1 Aluminum Salts

It aids in their gradual release at the injection site, which prolongs the immunological reaction. Research has demonstrated that aluminum adjuvants can either cause apoptosis or activate nucleotide binding domain

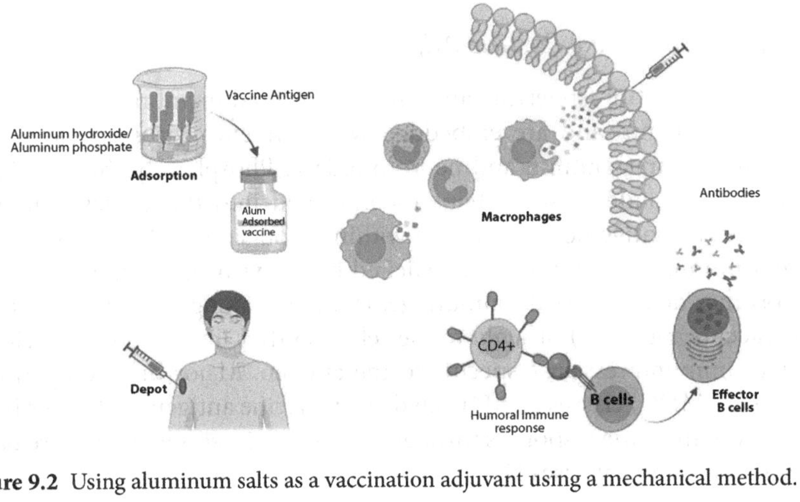

Figure 9.2 Using aluminum salts as a vaccination adjuvant using a mechanical method.

leucine rich containing family-pyrin domain containing- 3 inflammasomes (NLRP3), which releases harmful signals and ultimately activates the immune system. The mechanism is explained in Figure 9.2.

9.1.1.2 *Liposomal Adjuvants*

Adjuvant system AS01 (GSK): It has been founded on the synthesis of common liposomes, i.e., from cholesterol and dioleoylphosphatidylcholine, as well as two immunostimulants: the detoxified product MPL, which is made from Salmonella minnesota strain R595's lipopolysaccharide extract and QS21, a refined saponin that was extracted from the bark of Quillaja saponaria Molina, whereas cholesterol in liposomes can lessen QS21's hemolytic toxicity and increase the safety of adjuvants, MPL and QS21 in AS01 work in concert.

9.1.1.3 *Emulsion-Based Adjuvants*

Emulsion adjuvants can quickly develop significant humoral immunity. When used with emulsion adjuvants, children and the elderly (who often have weakened immune systems) respond better than when aluminum adjuvants are used. There are several different types of oil-in-water or water-in-oil emulsions in this family of adjuvants. Adjuvants containing water in oil, like complete Freund's adjuvants (CFA), are considered the most reactive substances and are thus not recommended for use by humans. Oil-in-water emulsions, which contain squalene-based adjuvants (MF59, AS02, and AS03), exhibit a reduced but still notable reactogenicity [5, 6].

9.1.1.4 *Virus-Like Particle Adjuvant*

Virosomes are tiny, spherical, and unilamellar lipid membrane vesicles that measure 150 nm and are embedded with influenza virus envelope proteins such hemagglutinin and neurominidase. Phosphatidylcholine bilayer liposomes include these proteins incorporated into them. Depending on whether the antigenic epitopes are within or outside of the virosome, different immune responses can be elicited by the virosome adjuvant system. Virosomes deliver a weak immunogenic antigen along with their contents, the vaccine antigen(s), straight to the cells that they intend to infect, triggering an immune response specific to the antigen. Major histocompatibility complex (MHC) classes I and II can deliver vaccine antigens to trigger B-cell and T-cell immune responses through the use of virosomes, which are particles that resemble viruses [7].

9.1.1.5 Saponin-Based Adjuvant

Saponins are the naturally occurring glycosides. Researchers are more interested in using them as a vaccine adjuvant due to their capacity of activation of the human immune system.

Novavax is the patent holder of "Matrix MTM," which is considered as a next-generation saponin-based adjuvant. It was demonstrated that Matrix MTM could activate lymphocytes, dendritic cells (DCs), and granulocytes in draining lymph nodes (dLNs), as well as increase the amount of cells in the spleen and dLNs. AS02, a malaria vaccine that combines MPL and QS-21, is now in the phase 2 trials [8].

Table 9.1 Adjuvanted vaccines currently licensed by FDA and EMA for human use. Licenced and novel vaccinations using adjuvants are as follows:

Trade name & manufacturing company	Adjuvant	Approval	Type & route of administration
ADACEL Mfg. By Sanofi Pasteur	AlPO4 (0.3mg AL3+)	By FDA on June 10, 2005	Tetanus toxoid, Reduced Diphtheria Toxoid and Acellular Pertussis Vaccine Intramuscular
AVAXIM Mfg. By Sanofi Pasteur	Al (OH)3 (0.3mg Al3+)	In United Kingdom on 1996	Inactivated hepatitis A vaccine Intramuscular
FENDRIX Mfg. By GSK	AS04 containing 3-o-dexacyl-40-monophosphoryl lipid A (MPL) 50 μg adsorbed on Al (OH)3 (0.5mg Al3+)	By EU on 2005	Hepatitis B (recombinant) vaccine Intramuscular
BEXSERO Mfg. By GSK	Al (OH)3 (0.5mg Al3+)	By The European Commission on January 14, 2013	Group B meningococcal vaccine (recombinant) Intramuscular

(*Continued*)

Table 9.1 Adjuvanted vaccines currently licensed by FDA and EMA for human use. Licenced and novel vaccinations using adjuvants are as follows: (*Continued*)

Trade name & manufacturing company	**Adjuvant**	**Approval**	**Type & route of administration**
BIOTHRAX Mfg. By Emergent Biosolutions	Al (OH)3 (0.6mg Al3+)	By FDA in 1970	Anthrax vaccine adsorbed Subcutaneous/ Intramuscular
BOOSTRIX Mfg. By GSK	Al (OH)3 (0.3mg Al3+) and AlPO4 (0.2 mg Al3+)	By FDA in May 2005	Diphtheria, Tetanus and Pertussis (acellular component) vaccine (reduced antigen content) Intramuscular
CERVARIX Mfg. By GSK	AS04 containing 3-o-dexacyl-4'-monophosphoryl lipid A (MPL) 50 µg adsorbed on Al (OH)3 (0.5mg Al3+)	In European Union in September 2007	Human papillomavirus vaccine (type 16,18) (recombinant) Intramuscular
FLUAD Mfg. By Sanofi Pasteur	MF 59 containing 9.75mg of squalene, 1.175mg of polysorbate 80, 1.175mg of sorbitan trioleate	By FDA in November 2015	Inactivated Influenza Vaccine, surface antigen Intramuscular
ENGERIX B Mfg. By GSK	Al (OH)3 (0.25mg Al3+)	By FDA in 1981	Hepatitis B (recombinant) vaccine Intramuscular
GARDASIL 9 Mfg. By Merck & Co	Al2 (PO4) (OH)(SO4) (0.5mg di Al3+)	By FDA in December 2014	Human papillomavirus 9-valent vaccine (recombinant) Intramuscular

(*Continued*)

Table 9.1 Adjuvanted vaccines currently licensed by FDA and EMA for human use. Licenced and novel vaccinations using adjuvants are as follows: (*Continued*)

Trade name & manufacturing company	**Adjuvant**	**Approval**	**Type & route of administration**
GARDASIL Mfg. By Merck & Co	Al2 (PO4) (OH)(SO4) (0.225mg Al3+)	By FDA in 2006	Human papillomavirus quadrivalent (types 6,11,16,18) vaccine (recombinant) Intramuscular
HBVAXPRO Mfg. By Merck & Co	Al2 (PO4) (OH)(SO4) (0.25mg Al3+)	By FDA in 1986	Hepatitis B vaccine, purified antigen (recombinant) Intramuscular
HAVRIX Mfg. By GSK	Al (OH)3 (0.5mg Al3+)	By FDA on 1995	Inactivated hepatitis A vaccine Intramuscular
IXIARO Mfg. By Valneva SE	Al (OH)3 (0.25mg Al3+)	By EMA in 2009	Japanese Encephalitis virus vaccine, inactivated, adsorbed Intramuscular
HEPLISAV- B Mfg. By Dynavax Technologies	CpG 1018	By FDA in 2017	Hepatitis B vaccine (recombinant) Intramuscular
NEISVAC-C Mfg. By Pfizer	Al (OH)3 (0.5mg Al3+)	In European union 2000s	Conjugated polysaccharide vaccine Intramuscular

(*Continued*)

Table 9.1 Adjuvanted vaccines currently licensed by FDA and EMA for human use. Licenced and novel vaccinations using adjuvants are as follows: (*Continued*)

Trade name & manufacturing company	**Adjuvant**	**Approval**	**Type & route of administration**
INFANRIX HEXA Mfg. By GSK	Al (OH)3 (0.5mg Al3+)	By EMA in October 2000	Diphtheria, tetanus, pertussis, (acellular component), anti-hepatitis B (recombinant), polio (inactivated) and anti-Haemophilus influenza type b conjugate vaccine Intramuscular
HEXYON Mfg. By Sanofi Pasteur	Al (OH)3 (0.6mg Al3+)	By EMA in April 2013	Diphtheria, tetanus, pertussis, (acellular component), hepatitis B (recombinant), poliomyelitis (inactivated) and Haemophilus influenzae type b conjugated vaccine influenza type b conjugated vaccine Intramuscular
IMOVAX TETANY Mfg. By Sanofi Pasteur	Al (OH)3 (0.6mg Al3+)	By FDA in 1991	Tetanus vaccine Intramuscular

(*Continued*)

Table 9.1 Adjuvanted vaccines currently licensed by FDA and EMA for human use. Licenced and novel vaccinations using adjuvants are as follows: (*Continued*)

Trade name & manufacturing company	**Adjuvant**	**Approval**	**Type & route of administration**
KINRIX Mfg. By GSK	Al (OH)3 (0.6mg Al3+)	By FDA in 2008	Diphtheria and tetanus toxoid and acellular pertussis adsorbed and inactivated poliovirus vaccine Intramuscular
TDVAX Mfg. By Sanofi Pasteur	AlPO4 (0.5mg AL3+)	By FDA in 2010	Tetanus & diphtheria toxoids adsorbed Intramuscular
SYNFLORIX Mfg. By GSK	AlPO4 (0.5mg AL3+)	By FDA in 2010	Pneumococcal polysaccharide conjugated vaccine Intramuscular
TRIAXIS POLIO Mfg. By Sanofi Pasteur	AlPO4 (0.33mg AL3+)	By FDA in 2000s	Diphtheria, tetanus, pertussis (acellular components) and poliomyelitis (inactivated) vaccine (reduced antigenic content) Intramuscular

(*Continued*)

Table 9.1 Adjuvanted vaccines currently licensed by FDA and EMA for human use. Licenced and novel vaccinations using adjuvants are as follows: (*Continued*)

Trade name & manufacturing company	Adjuvant	Approval	Type & route of administration
VAXELIS Mfg. By Sanofi Pasteur	AlPO4 (0.17mg AL3+), Al2 (PO4) (OH) (SO4) (0.15mg di Al3+)	By FDA on October 23, 2018	Diphtheria, tetanus, pertussis, (acellular component), hepatitis B (recombinant), poliomyelitis (inactivated) and Haemophilus influenzae type b vaccine Intramuscular
VAQTA Mfg. By Merck & Co	Al2 (PO4) (OH) (SO4) (0.45mg di Al3+)	By FDA on July 28, 1995	Inactivated Hepatitis A vaccine Intramuscular
SHINGRIX Mfg. By GSK	AS01B containing Quillaja Saponaria Molina plant extract, fraction 21(QS-21) 50µg, 3-O-dexacyl-40-monophosphoryl lipid A (MPL) from Salmonella Minnesota 50µg	By FDA in October 2017	Herpes zoster vaccine (recombinant) Intramuscular
RECOMBIVAX HB Mfg. By Merck & Co	Al2 (PO4) (OH) (SO4) (0.25-0.5mg Al3+)	By FDA in 1985	Hepatitis B vaccine (recombinant) Intramuscular

In Table 9.1, we can see that aluminum salts are used as an adjuvant in the great majority of vaccinations that are now approved for use in human body by the FDA and EMA bodies. It is important to emphasize that these adjuvants are the most ancient ones utilized in vaccine formulations and that increasing the quantity of novel adjuvants seems to be a must for enhancing vaccination safety and effectiveness [9].

9.2 Method and Mechanism of Action

Vaccine adjuvants are special extras added to vaccines. They boost and change the immune response. Their job is to make vaccines work better. They do this by making the immune system react stronger and longer. Different adjuvants work in different ways. But generally, they improve how vaccine parts are shown to immune cells. They also activate those cells and help make proteins that control immune reactions. A schematic representation of the onset and mechanism of action is explained in Figure 9.3.

- One way adjuvants work is by improving how vaccine parts are presented. They help cells like dendritic cells, macrophages, and B cells take in and show off the vaccine parts. This showing off process is key for triggering the adaptive immune response. That response includes making antibodies against the vaccine parts, plus activating special T cells.
- Forming antigen depots is another adjuvant function. A reservoir builds at the injection site. It steadily doles out antigens. This extends immune engagement duration. Fewer vaccine doses may be needed as a result.
- Forming antigen depot is another adjuvant function. A reservoir builds at the injection site. It steadily doles out antigens. This extends immune engagement duration. Fewer vaccine doses may be needed as a result [10].
- Adjuvants activate innateimmune receptors. They stimulate Toll-like receptors (TLRs), NOD-like receptors, and others. This triggers signaling pathways. These pathways lead to the production of proinflammatory cytokines, chemokines, and costimulatory molecules. This enhances the overall immune response to the vaccine.
- Adjuvants promote the maturation and activation of APCs. This leads to increased expression of costimulatory

molecules (e.g., CD80 and CD86). It also leads to increased expression of MHC molecules. This enhances the presentation of vaccine antigens to T cells. It results in a more robust adaptive immune response [11].

- The right combination of immune response types is crucial. Adjuvants sway this balance. Some boost interferon gamma and cellular defenses, like a Th1 boost. Others ramp up antibody production for a Th2 effect. The ideal response tackles different pathogens effectively [12].

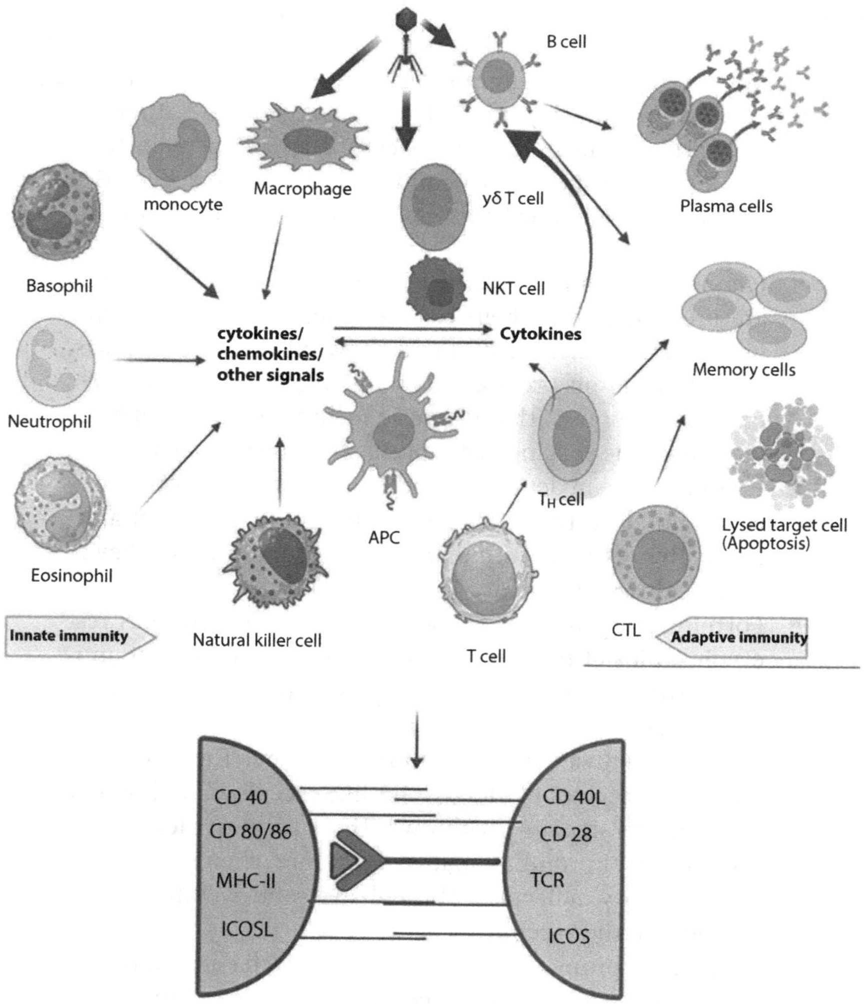

Figure 9.3 Innate and adaptive immune system mechanism of action.

- Activation of the Complement System: Some adjuvants can trigger the complement system, which is part of the innate immune response. This leads to improved phagocytosis, inflammation, and antigen presentation.
- Formation of Immunological Memory: Adjuvants help strengthen the development of protection over the long run against diseases. They can boost the creation of memory B cells and long-lasting plasma cells, as well as memory T cells. This ensures a quick and effective immune response when exposed to the pathogen again later on.

9.3 Approaches and Perceptions of Adjuvant Safety in Public Health

Adjuvants are special ingredients added to vaccines. They help boost the body's immune response to vaccine antigens. This makes vaccines more effective. Adjuvants can be organic or inorganic compounds. They can also be immune stimulating complexes or new formulations. Their goal is to enhance specific immune pathways. Some common adjuvants are aluminum salts like alum. Others are oil-in-water emulsions such as MF59 and AS03. Toll-like receptor (TLR) agonists are another type of adjuvant. But these adjuvants are safe. That is a key question. Safety testing is crucial for all adjuvants before they are approved for use in vaccines.

Many people worry about adjuvants when making vaccines. They wonder: Are adjuvants safe? To prove safety, scientists study adjuvants carefully. First, they run tests on animals. Then, people volunteer for clinical trials. After vaccines get approved, doctors monitor for side effects. Top health groups like the FDA, EMA, and WHO set rules about adjuvant safety testing. They aim to prevent harm. In short sentences and long sentences, the adjuvant testing process leaves no stone unturned. Researchers analyze the very possible risk. Public health experts would not approve adjuvants unless they prove completely safe [13].

9.3.1 Preclinical Safety Evaluation of Adjuvants

- Physicochemical Characterization: Before animal or human testing, adjuvants go through a deep study. Their makeup, stability, particle size, surface charge, and how they interact with antigens are determined. This helps find any impurities, clumping issues, or unexpected reactions [14].

- *In Vitro* Studies:
  - ➢ Cell Culture Models: Adjuvants are tested in lab dishes with cells. This looks as if they are toxic to cells, cause inflammation, make cells release cytokines, and activate cells. Different immune cells like macrophages, dendritic cells, and lymphocytes mimic immune responses [15].
  - ➢ Immunological Assays: Methods like enzyme linked immunosorbent assay (ELISA), flow cytometry, and gene tests evaluate adjuvant effects. For example, they check antigen presentation, cytokine release (like IL-1β and tumour necrosis factor-α (TNF-α)), and if markers show immune cell activation [16].
- Studies Over Animals:
  - ➢ Assess to acute toxicity: Obtain early safety profiles centered around systemic reactions on adjuvant administration, target organ, and dose dependence [17].
  - ➢ Periodic medication appraisals: Investigate the adjuvants' long-term and long-term consequences, taken into thought the immune-mediated response, variations the body creates, and long-term toxicity.
  - ➢ Reproductive and developmental toxicity studies: Assess the consequences on adjuvant exposure on mother's fertility, fetal development, and neonatal development.
- Immunogenicity assessment: Investigate adjuvant findings in terms of animal analogs of human vaccine response, notably antigen-specific immune responses, antibody generation (humoral immunity), T mobile activation, cell synthesis and memory, and an adjuvant's clinical safety evaluation.

9.3.2 Adjuvants' Clinical Safety Evaluation

9.3.2.1 Phase I Clinical Trials

- ➢ Studies on Dose Escalation: Set down safe dosage ratios and analyze all adjuvanted vaccines' initial safety, tolerability, and pharmacokinetics in healthy volunteers [18].
- ➢ Immunogenicity Assessment: In order to assess the therapeutic value of an adjuvant, measure the immune responses that a vaccination induces, antibody titers, and cellular immune reactions [19].

9.3.2.2 *II/III Phase Clinical Trials*

- Comprehensive Safety Study Methods: Investigate the safety, immunogenicity, and beneficial effects of adjuvanted boosters in a wide variety of populations, including as children, the elderly, and patients with impaired immune systems [20].
- Long-Term Follow-Up: Keep track of vaccine recipients for extended periods of time in order to identify immune-mediated reactions, uncommon side effects, and outcomes connected to vaccinations.

9.3.2.3 *Post-Marketing Surveillance*

- Pharmacovigilance Programs: Utilize active surveillance networks, spontaneous reporting systems, and adverse event tracking in healthcare settings to continuously monitor vaccination safety [21].
- Risk Communication: To foster trust and openness, inform the public, public health organization, and healthcare professional on the safety of vaccine [22].

9.4 Guidelines and Regulatory Considerations

- Regulatory Agencies:
  - FDA (Food and Drug Administration): Offers recommendations on adjuvanted vaccine requirements for preclinical and clinical testing, risk assessment, manufacturing standards, and post-licensure surveillance.
  - Based on scientific evidence, the European Medicines Agency (EMA) assesses adjuvanted vaccines for marketing authorization in Europe, placing a strong emphasis on efficacy, safety, and quality.
  - The World Health Organization (WHO) establishes international standards, guidelines, and recommendations for the development of vaccines, including evaluations of adjuvant safety and efficacy.

- Rules:
  - The International Conference on Harmonization (ICH) guidelines provide guidelines for risk management plans, pharmacovigilance, clinical trial design, and preclinical safety testing for vaccines and adjuvants.
  - ICH M3(R2): Particular recommendations for non-clinical safety investigations of pharmaceuticals generated from biotechnology, including adjuvants, covering evaluations of immunotoxicity, reproductive toxicity, and toxicology [23].

9.5 Adjuvant Safety Testing With Emerging Technologies

- Advanced Immunological Assays
  - High-Dimensional Flow Cytometry: After adjuvant treatment, characterize immune cell subsets, activation states, and functional responses in intricate immunological settings.
  - Transcriptomic Profiling: Find molecular fingerprints of adjuvant-induced immune responses and toxicity by examining gene expression patterns in immune cells, tissues, and organs [24].
- Regarding *In Silico* Modeling:
  - Informatized Toxicology: Utilize systems biology techniques, molecular docking simulations, and computational models to predict adjuvant interactions, toxicity pathways, and immune modulatory effects [25].
  - Structure–Activity Relationships: Assess the relationships between the adjuvant's structure and function, pinpoint the major molecular factors that determine safety and effectiveness, and create safer adjuvant variations [26].
- Technologies Omics:
  - Proteomics and Metabolomics: Define biochemical pathways and toxicity mechanisms by profiling protein expression, post-translational modifications, and metabolic alterations in response to adjuvants [27].

- ➢ Epigenomics: Examine epigenetic changes (such as DNA methylation and histone acetylation) brought about by adjuvants to better understand how immune activation relates to long-term immune memory and possible hazards [28].
- 3D Tissue Culture and Organoid Models:
  - ➢ Organ-on-a-Chip Platforms: High-throughput screening and personalized medicine techniques are made possible by these microfluidic devices that mimic human organ function, immunological interactions, and drug/adjuvant responses.
  - ➢ Precisely Cut Organ Slice Cultures: This *ex vivo* cultures preserve tissue architecture, immune cell populations, and functional responses, offering valuable information about the toxicity and adjuvant effects unique to certain organs [29].

9.6 Conclusion

The creation of immunity against microorganisms that cause illness is the ultimate aim of vaccination. Different immune responses are needed for protective immunization against various diseases, and these responses can be produced by utilizing the right vaccination adjuvant. Traditional alum-based formulations have given way to more sophisticated nanoparticle-based formulations in the field of vaccination adjuvants [30]. It contributes to the creation of new Pam3CSK4 analogs, such as 1,2,3-triazoles based on carbohydrate di- and tri-lipidated cysteine that show adjuvant qualities in vaccine formulations against cancer, viruses, and bacteria. Saponin mimics have been produced that improve the immune response by modifying the lipid and carbohydrate scaffolds (e.g., by adding a linker). But despite extensive study on a-GalCer components, which produced a large number of analogs and mimics, dismal results have been published in the literature. Given the variability of target epitopes and the ongoing emergence of new pathogens and tumour antigens, the research community aims to ensure a high level of broad protection by developing new concepts that are no longer based on structural vaccinology. Instead, new immunostimulants and innovative delivery systems that combine efficacy (with long-lasting memory immune responsiveness) and safety are being introduced.

References

1. Wu, Z. and Liu, K., Overview of vaccine adjuvants. *Med. Drug Discov.*, 11, 100103, 2021.
2. Rambe, D.S., Giudice, G.D., Rossi, S., Sanicas, M., Safety and Mechanism of Action of Licensed Vaccine Adjuvants. *Int. Curr. Pharm. J.*, 4, 8, 420–431, 2015.
3. Montomoli, E., Piccirella, S., Khadang, B., Mennitto, E., Camerini, R., De Rosa, A., Current adjuvants and new perspectives in vaccine formulation. *Expert Rev. Vaccines.*, 10, 7, 1053–1061, 2011.
4. Garçon, N., Leroux-Roels, G., Cheng, W.-F., Vaccine adjuvants. *Perspect. Vaccinol.*, 1, 1, 89–113, 2011.
5. Petrovsky, N., Comparative Safety of Vaccine Adjuvants: A Summary of Current Evidence and Future Needs. *Drug Saf.*, 38, 11, 1059–1074, 2015.
6. Wang, Z.-B. and Xu, J., Better Adjuvants for Better Vaccines: Progress in Adjuvant Delivery Systems, Modifications, and Adjuvant–Antigen Codelivery. *Vaccines*, 8, 1, 128, 2020.
7. Verma, S.K., Mahajan, P., Singh, N.K. *et al.*, New-age vaccine adjuvants, their development, and future perspective. *Front. Immunol.*, 14, 1043109, 2023.
8. Shi, S., Zhu, H., Xia, X., Liang, Z., Ma, X., Sun, B., Vaccine adjuvants: Understanding the structure and mechanism of adjuvanticity. *Vaccine*, 37, 24, 3167–3178, 2019.
9. Facciolà, A., Visalli, G., Laganà, A., Di Pietro, A., An Overview of Vaccine Adjuvants: Current Evidence and Future Perspectives. *Vaccines*, 10, 5, 819, 2022.
10. Awate, S., Babiuk, L.A., Mutwiri, G., Mechanisms of Action of Adjuvants. *Front. Immunol.*, 4, 1–10, 2013, [Internet] Available from: http://journal.frontiersin.org/article/10.3389/fimmu.2013.00114/abstract.
11. Morrow, M.P., Yan, J., Pankhong, P. *et al.*, Unique Th1/Th2 Phenotypes Induced during Priming and Memory Phases by Use of Interleukin-12 (IL-12) or IL-28B Vaccine Adjuvants in Rhesus Macaques. *Clin. Vaccine Immunol.*, 17, 10, 1493–1499, 2010.
12. Zhao, T., Cai, Y., Jiang, Y. *et al.*, Vaccine adjuvants: mechanisms and platforms. *Signal Transduct. Target. Ther.*, 8, 1, 283, 2023.
13. Moni, S.S., Abdelwahab, S.I., Jabeen, A. *et al.*, Advancements in Vaccine Adjuvants: The Journey from Alum to Nano Formulations. *Vaccines*, 11, 11, 1704, 2023.
14. Berbel Manaia, E., Paiva Abuçafy, M., Chiari-Andréo, B.G., Lallo Silva, B., Oshiro-Júnior, J.A., Chiavacci, L., Physicochemical characterization of drug nanocarriers. *Int. J. Nanomed.*, 12, 4991–5011, 2017.
15. Jannus, F., Medina-O'Donnell, M., Neubrand, V.E. *et al.*, Efficient *In Vitro* and *In Vivo* Anti-Inflammatory Activity of a Diamine-PEGylated Oleanolic Acid Derivative. *Int. J. Mol. Sci.*, 22, 15, 8158, 2021.

16. Alhajj, M., Zubair, M., Farhana, A., Enzyme Linked Immunosorbent Assay. StatPearls (Internet), 2023. https://www.ncbi.nlm.nih.gov/books/NBK555922/#__NBK555922_dtls__
17. Greaves, P., *Histopathology of preclinical toxicity studies: interpretation and relevance in drug safety studies*, 4th ed, Elsevier/AP, Amsterdam, Boston, 2012.
18. Fawole, O.A., Srivastava, T., Feemster, K.A., Student health administrator perspectives on college vaccine policy development and implementation. *Vaccine*, 37, 30, 4118–4123, 2019.
19. Schmidt, A.L., Zollo, F., Scala, A., Betsch, C., Quattrociocchi, W., Polarization of the vaccination debate on Facebook. *Vaccine*, 36, 25, 3606–3612, 2018.
20. Mulia, I.E., Ueda, N., Miyoshi, T., Gusman, A.R., Satake, K., Machine learning-based tsunami inundation prediction derived from offshore observations. *Nat. Commun.*, 13, 1, 5489, 2022.
21. Zacharia, A., Harberts, E., Valencia, S.M. *et al.*, Optimization of RG1-VLP vaccine performance in mice with novel TLR4 agonists. *Vaccine*, 39, 2, 292–302, 2021.
22. Larson, H.J., De Figueiredo, A., Xiahong, Z. *et al.*, The State of Vaccine Confidence 2016: Global Insights Through a 67-Country Survey. *EBioMedicine*, 12, 295–301, 2016.
23. Kille, J.W., Regulatory Toxicology, in: *A Comprehensive Guide to Toxicology in Preclinical Drug Development*, pp. 677–711, Elsevier, Amsterdam, Netherlands, 2013, [Internet] [cited 2024 Mar 14]Available from: https://linkinghub.elsevier.com/retrieve/pii/B9780123878151000289.
24. Wang, Y., Bi, Y., Chen, X. *et al.*, Histone Deacetylase SIRT1 Negatively Regulates the Differentiation of Interleukin-9-Producing CD4 + T Cells. *Immunity*, 44, 6, 1337–1349, 2016.
25. Gupta, S., Kapoor, P., Chaudhary, K. *et al.*, *In silico* approach for predicting toxicity of peptides and proteins. *PLoS One*, 8, 9, e73957, 2013.
26. Medina-Franco, J.L., Giulianotti, M.A., Welmaker, G.S., Houghten, R.A., Shifting from the single to the multitarget paradigm in drug discovery. *Drug Discov. Today*, 18, 9–10, 495–501, 2013.
27. Wishart, D.S., Emerging applications of metabolomics in drug discovery and precision medicine. *Nat. Rev. Drug Discov.*, 15, 7, 473–484, 2016.
28. Moosavi, A. and Motevalizadeh Ardekani, A., Role of Epigenetics in Biology and Human Diseases. *Gene*, 581, 1–13, 2016.
29. Proctor, W.R., Foster, A.J., Vogt, J. *et al.*, Utility of spherical human liver microtissues for prediction of clinical drug-induced liver injury. *Arch. Toxicol.*, 91, 8, 2849–2863, 2017.
30. Chavda, V.P., Vuppu, S., Bezbaruah, R. *et al.*, Phytochemical Loaded Nanovehicles of Biopolymer for Breast Cancer: A Systemic review. *Clin. Complement. Med. Pharmacol.*, 3, 100114, 2023.

10

Shortcomings of Current Adjuvants and Future Prospects

Pankti C. Balar[1], Vasso Apostolopoulos[2] and Vivek P. Chavda[3*]

[1]Pharmacy Section, L.M. College of Pharmacy, Ahmedabad, India
[2]School of Health and Biomedical Sciences, RMIT University, Melbourne VIC, Australia
[3]Department of Pharmaceutics and Pharmaceutical Technology, L.M. College of Pharmacy, Ahmedabad, India

Abstract
Adjuvants play a pivotal role in increasing vaccine efficacy by enhancing immune responses to antigens. This encompasses the entire spectrum of adjuvant technology, from design and development to regulatory considerations. While classic adjuvants like aluminum salts and oil-in-water emulsions remain at the forefront, innovative technologies like Toll-like receptor agonists, saponins, and nanoparticles are on the rise, heralding more potent vaccines. However, these adjuvants have challenges, including adverse reactions and variability in efficacy across populations. Hence, there is a pressing need for safer and non-inflammatory alternatives. Exciting advancements in nanotechnology, computational design, and personalized medicine hold promise for the future. Collaboration across academia, industry, and regulatory sectors is essential to address these hurdles. Ultimately, this effort strives to push forward research in developing innovative, efficient, and safe vaccine formulations to effectively address global health threats.

Keywords: Adjuvant limitations, immunogenicity challenges, toxicity and side effects, future adjuvant innovations, regulatory and development hurdles

**Corresponding author*: Vivek.chavda@lmcp.ac.in

Vivek P. Chavda and Vasso Apostolopoulos (eds.) Emerging Pathways of Vaccine Adjuvants: A Nonspecific Stimulant of the Immune System, (255–262) © 2025 Scrivener Publishing LLC

10.1 Introduction

Adjuvants enhance the ability of antigens to stimulate potent, enduring immune responses using a lower concentration of antigen and hence reducing vaccine production expenses. Adjuvants have gained significant interest in recent years due to the advancement of purified, subunit, and synthetic vaccines [1]. Subunit protein–based and peptide-based vaccines have limited ability to stimulate robust immune responses, making adjuvants necessary to elicit an immunological response [2, 3]. Adjuvants can modulate the immune response to Th1 pro- or Th2 anti-inflammatory, which is crucial for overcoming diseases caused by intracellular pathogens such viruses, parasites, and bacteria [4]. Adjuvants are classified into two categories: immunostimulants and delivery systems. Immunostimulants act as danger signals, prompting antigen-presenting cells (APCs) to mature and become activated. This is achieved via their interaction with Toll-like receptors (TLRs) and other pattern recognition receptors, which then stimulate the production of antigen signals and co-stimulatory signals. This effectively boosts adaptive immune responses. Delivery systems facilitate antigen presentation by extending the presence of antigens in the body and by directing them toward lymph nodes or APCs [5]. The selection of an appropriate adjuvant to enhance immune responses is a major consideration in the development of vaccines. The immune system's response to a specific vaccine and adjuvant combination largely depends on individual factors, making it clear that no single adjuvant fits all antigens.

10.2 Limitations

The development of mRNA-based vaccinations has highlighted a second scientific obstacle, namely, the lack of knowledge regarding the effectiveness and appropriateness of adjuvants for nucleic acid–based vaccines. Conventional adjuvants often disrupt the expression of antigens from nucleic acid vaccines, affecting the immune response both in terms of quantity and quality. This interference is a result of the stimulation of antiviral pathways that aim to inhibit viral reproduction. The lack of data from adjuvant mRNA-based vaccines now hinders the development of enhanced mRNA formulations. However, this also presents a promising research opportunity. Given the high level of interest in the mRNA vaccine platform, obtaining this information is of utmost importance [6].

Despite being used as an adjuvant in humans for nearly a century and administered to billions of individuals, there are still unresolved concerns about the potential link between aluminum and illnesses like monomethyl fumarate (MMF) or Alzheimer's disease [7]. Although it is impossible for a vaccination to be completely risk-free, the likelihood of experiencing immediate severe adverse reactions from modern vaccines is quite low. Therefore, individuals can have confidence in the safety of the vaccines that are currently accessible. In order to promote the use of new adjuvants, it is crucial to enhance awareness and understanding of the evaluation of risks and benefits associated with vaccines. Due to the crucial significance of adjuvants in contemporary vaccinations, further resources are required to facilitate research aimed at gaining a more comprehensive comprehension of adjuvant mechanisms and their potential correlation with adjuvant toxicity. There is a growing demand for adjuvants that can improve vaccine efficacy without inducing unwanted side effects. An important research area to explore is the development of non-inflammatory adjuvants capable of enhancing vaccine potency without causing adverse effects or compromising safety [8]. There are still several obstacles in the advancement of oral adjuvants. As previously stated, the gastrointestinal tract is a challenging and demanding environment, causing instability in adjuvants, particularly biogenic adjuvants, and impeding the interaction between oral adjuvants and cells in the intestinal lining. Conversely, the small percentage of M cells in the intestinal epithelium would restrict the impact of M cell-mediated adjuvants. Nevertheless, it is anticipated that an increasing number of oral adjuvants, such as RANKL, capable of stimulating M-cell differentiation, will arise in the future to address this predicament. Another problem arises from the possible safety risks associated with adjuvants, particularly cytokine-derived adjuvants. Further evaluation is required to assess the efficacy, pharmacokinetics, and nanotoxicology of certain possible oral adjuvants [9, 10].

10.3 Advancements

The advancement of potent vaccinations is an element of public health, and the utilization of adjuvants has surfaced as a viable approach to augment vaccine effectiveness. Adjuvants boost the immune response to an antigen, therefore enhancing the effectiveness of vaccines. Adjuvants have been increasingly popular in recent years because they have the potential to decrease the amount of antigen needed, enhance the length of protection, and expand the immune response [11, 12]. Some adjuvants include the following:

a. Aluminum salts are widely used as adjuvants in vaccines, with a history of over 90 years. Studies have demonstrated that aluminum salts can boost the immune response to many antigens, including those used in vaccines for hepatitis B, human papillomavirus, and pneumococcal diseases. Nevertheless, the precise mechanism of action of aluminum salts remains incompletely comprehended, and there exist apprehensions regarding their safety and capacity to induce unfavorable reactions [4].
b. Oil-in-water emulsions, including MF59 and AS03, are a type of adjuvant that has received significant interest in recent years. Studies have demonstrated that these adjuvants effectively boost the immune response to several antigens and have been used in influenza and hepatitis B vaccines. The oil-in-water emulsions are believed to work by attracting immune cells to the injection site and improving the presentation of antigens to the immune system [13].
c. Additional categories of adjuvants currently under investigation include TLR agonists, saponins, and nanoparticles. Toll-like receptor agonists stimulate TLRs, which are crucial elements of the innate immune system. Saponins are innate substances that have demonstrated the ability to augment the immune response to several antigens, including those employed in HIV and malaria vaccines. Nanoparticles are submicron-sized particles, measuring less than 100 nm, which have demonstrated the ability to augment the immune response to antigens, including those used in cancer and infectious disease vaccines [14, 15].

Adjuvants in vaccines offer several benefits, such as reducing the dose of antigen, and enhance the duration of immune stimulation. Nevertheless, the utilization of adjuvants presents various obstacles, such as adverse reactions, and, as such, the requirement for careful formulation and quality control and the necessity for additional research to understand their mechanism of action [16].

10.4 This Book

Adjuvants serve as vital components in vaccines, enhancing the immune responses initiated by antigens and enhancing vaccine efficacy. This comprehensive book describes cutting-edge advancements in adjuvant technology, aiming to propel vaccine research toward developing safer and more potent vaccines to overcome global health challenges effectively. It offers information of vaccine adjuvants, encompassing their design, development, application, and their pivotal role in augmenting vaccine efficacy. The book also presents innovative strategies for adjuvant formulation, detailed characterization methods ensuring quality, and their significant contributions to licensed vaccines. Emerging technologies, such as nanomaterial-based adjuvants and non-invasive delivery methods, are also described as promising avenues for future vaccine development. These innovations hold the potential to revolutionize adjuvant design by providing tailored solutions to specific antigens. In addition, the book addresses critical regulatory guidelines, patent considerations, and safety concerns associated with adjuvants. It provides insightful discussions on the current limitations of adjuvants, highlighting the imperative need for safer alternatives that do not induce adverse reactions or exhibit variable efficacy across populations.

10.5 The Future

The future of adjuvant research is promising, with a strong focus on developing non-inflammatory adjuvants that can enhance vaccine effectiveness without compromising safety. Advances in nanotechnology, computational design, and personalized medicine are paving the way for developing adjuvants that are finely tuned to specific antigens and population needs. Additionally, a deeper understanding of immune mechanisms and biomaterials is anticipated to lead to the development of targeted and efficient adjuvants. Collaborative efforts between academia, industry, and regulatory bodies are deemed essential in overcoming current challenges and steering the field toward the creation of safer and more effective vaccine formulations. By integrating diverse expertise and fostering interdisciplinary collaborations, the book emphasizes the collective responsibility of the scientific community in advancing vaccinology. Ultimately, this holistic approach aims to address the current limitations of adjuvants, driving the field toward innovative solutions that enhance vaccine efficacy while ensuring safety and accessibility for all.

References

1. Jetha, K.L., Gogoi, N.R., Kumar, A. *et al.*, Chapter 11 - Parenteral vaccine delivery: From basic principles to new developments, in: *Advanced Vaccination Technologies for Infectious and Chronic Diseases*, V.P. Chavda, L.K. Vora, V. Apostolopoulos (Eds.), pp. 167–205, Academic Press, Cambridge, USA, 2024, [Internet] Available from: https://www.sciencedirect.com/science/article/pii/B9780443185649000047.
2. Apostolopoulos, V. and Chavda, V.P., Subunit protein-based vaccines, in: *Advanced Vaccination Technologies for Infectious and Chronic Diseases*, pp. 51–62, Elsevier, Cambridge, USA, 2024, [Internet] [cited 2024 Apr 24]. Available from: https://linkinghub.elsevier.com/retrieve/pii/B9780443185649000114.
3. Chavda, V.P., Ghali, E.N.H.K., Balar, P.C. *et al.*, Protein subunit vaccines: Promising frontiers against COVID-19. *J. Control. Release*, 366, 761–782, 2024.
4. Gupta, R.K. and Siber, G.R., Adjuvants for human vaccines—current status, problems and future prospects. *Vaccine*, 13, 14, 1263–1276, 1995, [Internet] Available from: https://www.sciencedirect.com/science/article/pii/0264410X9500011O.
5. Zhao, T., Cai, Y., Jiang, Y. *et al.*, Vaccine adjuvants: mechanisms and platforms. *Signal Transduct. Target. Ther.*, 8, 1, 283, 2023, [Internet] Available from: https://doi.org/10.1038/s41392-023-01557-7.
6. Singleton, K.L., Joffe, A., Leitner, W.W., Review: Current trends, challenges, and success stories in adjuvant research. *Front. Immunol.*, 14, 1105655, 2023, [Internet] Available from: https://www.frontiersin.org/journals/immunology/articles/10.3389/fimmu.2023.1105655.
7. Chavda, V.P., Gogoi, N.R., Shah, Y., Shah, D., Mazumder, B., Chapter 21 - New approaches to vaccines for autoimmunity, in: *Advanced Vaccination Technologies for Infectious and Chronic Diseases*, V.P. Chavda, L.K. Vora, V. Apostolopoulos (Eds.), pp. 391–414, Academic Press, Cambridge, USA, 2024, [Internet] Available from: https://www.sciencedirect.com/science/article/pii/B9780443185649000230.
8. Petrovsky, N., Comparative Safety of Vaccine Adjuvants: A Summary of Current Evidence and Future Needs. *Drug Saf.*, 38, 1059–1074, 2015.
9. Ou, B., Yang, Y., Lv, H., Lin, X., Zhang, M., Current Progress and Challenges in the Study of Adjuvants for Oral Vaccines. *BioDrugs*, 37, 2, 143–180, 2023, [Internet] Available from: https://doi.org/10.1007/s40259-022-00575-1.

10. Newsted, D., Fallahi, F., Golshani, A., Azizi, A., Advances and challenges in mucosal adjuvant technology. *Vaccine*, 33, 21, 2399–2405, 2015.
11. Facciolà, A., Visalli, G., Laganà, A., Di Pietro, A., An Overview of Vaccine Adjuvants: Current Evidence and Future Perspectives. *Vaccines*, 10, 5, 819, 2022.
12. Chavda, V.P., Acharya, D., Malla, S., Ashby, C.R., Tiwari, A.K., Chapter 20 - New approaches to vaccines for cancer, in: *Advanced Vaccination Technologies for Infectious and Chronic Diseases*, V.P. Chavda, L.K. Vora, V. Apostolopoulos (Eds.), pp. 357–390, Academic Press, Cambridge, USA, 2024, [Internet] Available from: https://www.sciencedirect.com/science/article/pii/B9780443185649000096.
13. Verma, S.K., Mahajan, P., Singh, N.K. *et al.*, New-age vaccine adjuvants, their development, and future perspective. *Front. Immunol.*, 14, 1043109, 2023, [Internet] Available from: https://www.frontiersin.org/journals/immunology/articles/10.3389/fimmu.2023.1043109.
14. Filipić, B., Pantelić, I., Nikolić, I. *et al.*, Nanoparticle-Based Adjuvants and Delivery Systems for Modern Vaccines. *Vaccines*, 11, 7, 1172, 2023.
15. Vasudevan, S.S., Kandrikar, T.Y., Sayyed, A.A., V, *et al.*, Chapter 25 - Nanoparticle-based vaccines and future vaccine technologies, in: *Advanced Vaccination Technologies for Infectious and Chronic Diseases*, V.P. Chavda, L.K. Vora, Apostolopoulos (Eds.), pp. 477–495, Academic Press, Cambridge, USA, 2024, [Internet] Available from: https://www.sciencedirect.com/science/article/pii/B9780443185649000023.
16. Chatzikleanthous, D., O'Hagan, D.T., Adamo, R., Lipid-Based Nanoparticles for Delivery of Vaccine Adjuvants and Antigens: Toward Multicomponent Vaccines. *Mol. Pharmaceutics*, 18, 8, 2867–2888, 2021, [Internet] Available from: https://doi.org/10.1021/acs.molpharmaceut.1c00447.

Index

Printed in the USA/Agawam, MA
September 16, 2026

910661.068